Printers Face Automation

Printers Face Automation

The Impact of Technology on Work and Retirement among Skilled Craftsmen

Theresa F. Rogers
Nathalie S. Friedman
Columbia University and
Barnard College

LexingtonBooks
D.C. Heath and Company
Lexington, Massachusetts
Toronto

Library of Congress Cataloging in Publication Data

Rogers, Theresa F
Printers face automation.

Bibliography: p.
Includes index.
1. Printers—United States. 2. Printing industry—Automation—Social aspects—United States. 3. Printers—United States—Retirement. 4. Machinery in industry—United States. I. Friedman, Nathalie S, joint author. II. Title.
Z243.U5R63 686.2'092'2 79-3047
ISBN 0-669-03310-3

Published simultaneously in Canada.

Printed in the United States of America.

International Standard Book Number: 0-669-03310-3

Library of Congress Catalog Card Number: 79-3047

Contents

List of Figure and Tables

Figure

Tables

Foreword

"You can't stop progress." Ever since the Luddites went on their machine-smashing binge in Britain's textile mills at the dawn of the industrial revolution, that has been society's admonition to workers and their unions whenever they resisted the introduction of technology that would make it possible to do more work with fewer people.

From the standpoint of the general welfare, there is much to be said for such advice. It may be argued that the exaltation of technological progress has exacted too high a price in corruption of the environment, obsession with gadgetry, and risk of thermonuclear annihilation. Yet it is indisputable that standards of living for all classes of citizens have been raised by the vast expansion in the quantity of goods and services that increased industrial efficiency and scientific inventiveness have permitted.

Indisputably, too, advanced methods have taken much of the drudgery out of the workplace and have held down production costs in ways of immeasurable benefit to consumers. Famine would be worldwide without the fantastic productivity that mechanization has brought to the farms. In every field from health to space travel, man's reach has been extended spectacularly by the onrush of science. Nowhere has this been truer than in the United States, with its creativity, its resources, and its dedication to the democratic ideal of upward mobility.

What has been lacking until recently, however, has been any discernible concern by anyone except the victims for the often devastating damage that technological change inflicts on workers whose jobs are wiped out or whose painfully acquired skills are rendered valueless in the name of progress. The dramatic new dimension given to the pace of change by automation and all the other applications of computers, lasers, microcircuitry, and systems engineering has now forced a recognition that inventiveness is just as necessary in guarding against human wreckage as in finding ways to capitalize on potential breakthroughs in the diversity and efficiency of industry.

One of the most comprehensive labor-management responses yet made to this need was the automation pact negotiated in 1974 by New York City's daily newspapers and their unionized printers. The pact ended a decade-long veto by the printers over use by the publishers of processes that represented a revolution in typesetting and composition, and it did so on a basis that gives both sides full protection on the issues each considered paramount.

For management the agreement provided absolute freedom to install new technology, flexibility in assigning personnel, and abolition of wasteful work rules that had been sacrosanct for a half-century. The printers, proud of both their skills as craftsmen and the high level of democracy that distinguishes their union, received promises of lifetime job security, plus emoluments and safeguards of a kind not enjoyed by any of the papers' other employees, union or executive. So sweeping were the new protections against permanently discharging a printer, even for disciplinary infractions, that the union's president could boast, "We have abolished industrial capital punishment."

In this book Professors Theresa F. Rogers and Nathalie S. Friedman undertake a fascinating exploration into the manner in which the printers and their union, cushioned by this precedent-setting accord, adjusted to a change that made obsolete almost all the skills they had spent their working lives perfecting and that stripped the union of the iron control it had long held over composing-room practices.

Much of what they found is highly reassuring, especially to those who worry about the supposed demise of the work ethic and who recognize that the adaptability of U.S. labor as well as U.S. industry will be a vital determinant of this country's ability to remain economically strong in a world of rapid technological innovation and increasingly fierce trade competition.

Despite the fact that most of the printers were well advanced in age, they responded eagerly to the opportunities that the publishers made available to acquire the lesser skills required to set type by computer and make up newspaper pages and advertisements through a process one step removed from cutting and pasting paper dolls. Far from feeling degraded by this devaluation of typographical artistry, most took pride in demonstrating the fantastic capabilities of the machines they had in their custody, if not in their command.

If that were the end as well as the beginning of the transformation, we could all feel just fine about the marvelous effectiveness of collective bargaining as a vehicle for producing a meeting of minds that worked out beneficially for everyone, including the readers, who found their dailies less pockmarked with typographical errors and (at least theoretically) could get more intensive coverage of the news through the publishers' ability to apply to the pursuit of editorial excellence money once expended on mechanical payroll.

Unfortunately, things became less idyllic for the workers in the composing room when the introduction of one step after another in new technology—and its ultimate potential is not yet realized—caused the need for printers to plummet at a much faster rate than they chose to retire even when management offered bonuses of as much as three years' pay to supplement the pensions and other incentives for early retirement. Printers could adapt to any aspect of change except total idleness. That was galling even at guaranteed pay in excess of $20,000 a year.

Automation in other departments had, for all practical purposes, ruled out the possibility of transfer to duties elsewhere in the paper. Inflation, eroding the purchasing power of every retired person's nest egg at a distressing rate, had made even the most generous pension and buy-out arrangements illusory unless they carried a cost-of-living escalator to make them inflationproof—an adjustment of great actuarial peril to the publishers. But staying on the job also lost its allure when there literally was nothing for the printers to do except tidy up, play cards, or chat.

For their union the long-term prospect is equally cheerless. Despite the resourcefulness its leaders have exhibited in trying to find new functions and services, the outlook inescapably is for obliteration of usefulness, thus wiping out an organization that is among the oldest and most respected in U.S. labor. The great virtue of the Rogers-Friedman study is the insight it gives not only into the immediate virtues of a well-conceived formula for dealing with the human consequences of automation but also into the longer-view implications for workers and unions of their acceptance of the doctrine that you cannot stop progress. Their findings make it impossible for all interested in a productive society to ignore the great distance we have yet to go before we achieve an ideal blend of efficiency and equity in the public interest.

A.H. Raskin
Associate Director,
National News Council
Formerly chief labor
correspondent, *The New York Times*

Acknowledgments

This book began as a research proposal submitted to the Labor Management Services Administration of the U.S. Department of Labor. The proposal was funded (Contract No. J-9-P-6-0020), and we wish to acknowledge our appreciation for this financial support and for the encouragement and interest of our first project officer, Herbert Lahne, and his successor, Leon Lunden. All contract research should have such informed and helpful project officers!

Until its close in June 1977, we were staff members of the Bureau of Applied Social Research, Columbia University, and we want to thank, in particular, four long-time Bureau friends and colleagues for their help: Allen Barton, Phyllis Sheridan, Pnina Grinberg, and Madeline Simonson. To this list we add Chris Kelly, the administrator of the Graduate School of Business, with which we are now affiliated. She willingly and ably helped us stretch the few remaining dollars of our government contract.

We also want to acknowledge the help of the following persons: our student research assistants, Amy Bernstein, Sue Erichsen, Irit Laor-Sherman, and Andrew Rak; and the staff of seven who assisted us in the interviews with printers on weekdays and weekends, early in the morning and late at night, Jill Bernstein, Suzanne Bernstein, Ann Haney, Nava Harlow, Miriam Kahn, Ann Miles, and Elaine Tollstrup.

The sheer number of printers and their wives, production managers, foremen, assistant foremen, staff, and officers of the union who helped us is so great that we cannot thank them individually in writing. We do, however, want to single out several whose support and assistance truly made the research possible:

Bertram Powers, president of Local 6, who thought it important to conduct the study and who graciously and promptly asked the chapel chairman at each paper to help us in every way possible.

The chapel chairmen (in 1976): Rocky Solomon at the (New York) *Daily News*, Joseph Di Napoli at the *New York Post*, and Bobby Cox at *The New York Times*, each of whom spent hours briefing us on union and shop matters.

The general foreman: Dan Schlicht at the *Daily News*, Rudy Rossiello at the *New York Post*, and John Fornazer at *The New York Times*, who gave us their full cooperation and repeat interviews and expedited our interviews with printers.

Arthur Wasserman and Gunnar Janger at Local 6, who gave us access to the statistical data we needed.

One informant merits special thanks. Carl Schlesinger, a printer at *The New York Times* and coauthor of *Union Printers and Controlled Automation* (Kelber and Schlesinger 1967), served as our chief technical consultant. He carefully explained the old and new methods of printing, answered our many questions, and offered valuable suggestions for the revision of an early draft of the manuscript.

Several colleagues—Abe Jaffe, Jim Kuhn, Loya Metzger, and Ann Pasanella—reviewed sections of the manuscript on short notice, always willingly. Their penciled comments were of great help both in tightening the manuscript and in encouraging us to complete it.

We want to express our special appreciation to two friends and colleagues, Jill Bernstein and Ken Lenihan. Each, in different ways, was a source of ideas and support throughout the long stretch of time it took to turn a file cabinet full of research materials into a book. Without their help, this effort would have been far more difficult and lonely. We also want to thank Charlotte Fisher for her expert typing of early draft material and Estelle Cooper for typing the final manuscript with great skill and, equally important, with good cheer and amazing tolerance for our last-minute changes.

On a more personal note, our husbands, Israel Friedman and David Rogers, are to be thanked for reading and commenting on the manuscript chapter by chapter and for making it easier in so many ways for us to make the book a reality.

At the same time that we acknowledge the considerable help we have received, we assume full and joint responsibility for the analysis and interpretations set forth. We stress the collaborative aspect of this book, because there is no senior author. The order of the names was decided by the toss of a coin.

Introduction

This book is about printers, men who have mastered a skilled craft and who belong to an old and distinguished union, the International Typographical Union (ITU). They are members of Local 6, also known as Big Six, and they are either working for or recently retired from one of New York City's three major newspapers, the *Daily News*, the *New York Post*, or *The New York Times*.

They chose a prestigious craft as their life's work, served a long apprenticeship, and finally became journeymen printers. They are walking encyclopedias about the requirements of printing—they can set intricate type, artistically lay out advertisements, proof their work, and repair the machinery they use. Accomplished and respected for their knowledge and skill, they have often been referred to as "the elite of blue-collar workers."

In 1976, there were 1700 of them at the *Daily News, New York Post*, and *The New York Times*. Their average age was fifty-six, with the youngest in his early thirties and the oldest a vigorous eighty-two. Many have worked thirty, forty, even more than fifty years, standing at long tables or seated before machinery three times their size, as they composed the printed page. Until recently, they wore large aprons to protect their clothing from the heavy stains of printer's ink, the sweetest smell on earth to them. They learned to live with the noise and clatter of massive machinery, to speed frantically to meet an edition deadline, to correct copy as it was rushed from the newsroom, and to repair hundred-year-old lumbering linotypes.

Now automation has transformed their workplace, the composing room. They can come to work in shirts and ties; they sit at small desks and lightly touch the keys of IBM Selectric typewriters, scan video display screens, or monitor mammoth electronic terminals. They can talk all they want about the craft that they mastered with such care, but the message is "Forget the old days; they are gone forever."

The technological revolution in the composing room is the result of an eleven-year collective-bargaining agreement signed in 1974 between the Publishers' Association of New York City, representing the *Daily News, New York Post*, and *The New York Times*, and Local 6. By 1974 almost every daily in the nation had already switched to automated typesetting; but New York is a strong union town, and for many years Local 6 could veto the introduction of new equipment. By 1974, however, the publishers were determined to automate the papers, and they negotiated a landmark agreement with the printers' union. The agreement gave management full freedom to automate the composing room. In return, printers won job security, a six-month paid productivity leave, periodic cost-of-living increases, and a dollar incentive to retire. The union also agreed that henceforth no new printers would be hired at the papers.

This book is concerned with the first three years of the contract and answers the question of what happened to the workers caught in this technological change. It reports their response to automation and to voluntary retirement spurred by a dollar incentive, as well as the impact of automation on their union and the changing direction of labor-management relations. Although the focus is one occupational group in one industry and in one city, the implications are relevant for other workers facing change at the workplace, for women and men considering retirement, and for a better understanding of some of the pressures on trade unions in the late twentieth century.

This book shows what it means for middle-aged and older men to come to terms with gigantic social change. It examines the meaning of work and of retirement under unusual circumstances, namely, lifetime job security and voluntary retirement. Technically, no one is forced out, and those who retire make the decision in light of what they consider their best interests. Under these conditions, what decisions and adjustments do men make and what are the consequences for their union, their employer, and their own lives?

The book is based on interviews with 408 printers, 87 retirees, and the wives of 50 retirees, as well as interviews with various officers of Local 6, production managers, foremen, and assistant foremen at the papers. We also interviewed 150 wives of working printers, but their views about work and retirement, as well as those of the few women printers at these papers, have been reserved for a separate publication.

We visited the composing room of each paper many times to learn how the new "cold type" was displacing the old hot-metal process and to observe the work being performed and its pace on all three shifts. Further, we attended chapel (shop) meetings at the papers, meetings of pensioners, and meetings of Local 6. Most of this work was done in 1976, but it has been supplemented by return visits and interviews between 1977 and 1979.

In the spring of 1976, when we began our interviews with printers, about half had been trained to operate the new equipment, and the other half were setting type by hand and repairing the old linotypes. Chapters 1 through 3 compare the men who had been retrained with the ones who had not. They explain how both types of men are coming to terms with the changeover and why some find the transition more difficult than others.

Chapters 4 and 5 focus on retirement. Chapter 4 juxtaposes the men who took the retirement incentive provided by the contract with those who rejected it in favor of working. Chapter 5 describes the life of the retired printer and his wife two years into retirement, how they appraise their decision, and the suggestions they offer the next generation of retirees.

As older workers retire and no new ones replace them, the imbalance between the working and the retired creates a severe problem for the union. Chapter 6 examines this problem and analyzes the steps that Local 6 is taking to meet the changing needs of its membership and, above all, to continue as a viable entity.

Because by 1979 the transition from hot metal to cold type was so complete in the composing room of the papers, we have added an epilogue, chapter 7, to this book. Its title, "Ghost Town," is self-explanatory. The old machinery is gone or shoved into a corner until it can be sold. Although in the past three years the workforce has declined from 1,700 to 1,300 printers, more must leave because productivity is so great the new way that there is not enough work to go around. Many middle-aged men are caught in a double bind: boredom and friction if they remain; limited skills and only a high school education to sell to another employer.

Chapter 8 presents the conclusions and implications of this research and places it in the context of policy issues now being faced by government, industry, unions, and the individual. Can older workers be retrained? What evidence is there of alienation from work and what does this portend? What are the sources of a worker's job satisfaction and what are his loyalties to his union and to his employer? What happens to the worker replaced by a machine? Then, what does it take for a man to retire? Who is most likely to choose this alternative and with what results?

On balance, the response of printers to automation is cause for optimism. These men have shown the capacity to change, to learn the new methods, and to accept the loss of their craft. But the tragedy is that they have not been sufficiently prepared for the transition away from printing.

A Technical Note

This book was written by sociologists, and although we hope our colleagues will find it useful, we also intend it for a wider audience. For this reason most of the quantitative data are appended to each chapter rather than presented in the text. The specialized reader also will find a detailed account in appendix A of the research methods and sampling procedures employed, and in appendix B of the distribution of responses to the survey items asked working and retired printers, as well as their wives.

The printers whom we have studied represent a finite universe, which, under the impetus of technological change, is rapidly shrinking and soon will disappear. Our sample of working printers consisted of about one-fourth of this universe in 1976; the retired printers represented about two-thirds of the universe. In estimating the sampling error, that is, the extent to which the findings can be generalized to the universe of printers, we have used the finite multiplier for modifying the sampling error as determined for an infinite universe. Explanation is given in many statistical texts; we used Richmond (1964).

Finally, in the data analysis we have been less concerned with generalizing to the finite universe of printers than with explaining their differential response to the conditions that they confront. All the same, some readers may want to know which of our findings are statistically significant. With few exceptions (for example, when the lack of a difference is itself an important substantive finding) we have discussed only those findings which are significant at the .05 level or higher. The one remaining exception to this convention is that when a single percentage difference was not itself significant but was part of a series of related items, all pointing in the same direction, we have reported this pattern.

Printers Face Automation

1 The Old Technology and the New

Introduction

The increasing prosperity and security of the U.S. workingman and his family is perhaps the major success story of the decades since World War II (Rainwater 1971). With this rise in the material standard of living, interest has turned to the *quality* of people's lives, in particular, to the quality of working life. A person's job, after all, occupies the largest block of his time; in fact, two-thirds of the average man's waking life is spent making a living. And with more and more women entering the work force, an ever-increasing proportion of the adult population is working.

Small wonder, then, that in the last decade the quality of working life has emerged as a salient issue. In 1973 a report to the Secretary of Health, Education, and Welfare showed job dissatisfaction at all levels of the workforce and called for concerted efforts to humanize work (*Work in America* 1973). In 1974 an American Assembly conference at Arden House explored the problems of the worker stemming from a rapidly changing society (Rosow 1974). In that same year, Braverman published his analysis of the labor process in modern capitalist society, *Labor and Monopoly Capital*, subtitled *The Degradation of Work in the Twentieth Century.*

Underlying this attention to the quality of working life is the concern that technological change or automation is dehumanizing the workplace and producing low morale, absenteeism, and alienation. As a result, complains the Department of Health, Education, and Welfare (HEW) report, more than 10 million U.S. workers are dissatisfied with their jobs (*Work in America* 1973).

In other words, it is almost a given that automation and job satisfaction are inextricably linked. And the direction of the linkage is taken for granted: Automation means lowered job satisfaction. The reasoning is that the introduction of a complex technology means that individual skill and autonomy over the work process give way to a machine and that craftsmanship acquired over long years becomes obsolete.

This book examines the impact of a dramatic technological change which has taken place in one of the oldest, most respected trades in the United States—printing. Some fifteen years ago, one writer, Blauner, described the printer as "almost the prototype of the non-alienated worker in modern industry." At the same time, he added (Blauner 1964):

> Yet it is not certain how long printers can maintain this position, for technological innovations and economic developments threaten to eliminate not only the typesetter's control but the job itself. The newspaper industry has developed a process by which printed type can be set automatically.

These days, typesetting *is* largely automatic. By 1976 the printed type required for advertising, classifieds, and obituaries, as well as most of the movie, theater, and TV listings, was being set at the *Daily News* and *The New York Times* by the new electronic "cold-type" method rather than by the old hot-metal process. As a result, the workday of a large proportion of the printers is very different from what it used to be.

From the Old Technology to the New

Until the early 1960s, when the first teletypesetter was introduced into *The New York Times* composing room (the printers' workplace), printing had remained virtually unchanged. As one printer put it, "Mr. Gutenberg would have been at home in the composing room as recently as twenty years ago, but no more."

For a short time, the clatter of the old linotype machines will still be heard because about one-half of the composing room force is helping to turn out the newspaper each day in much the same way as their fathers and grandfathers before them. As then, the work of these men in the composing room begins as edited stories from the newsroom are passed on to a linotype operator for typesetting.

The linotype machine is a large, lumbering piece of equipment reminiscent of a Rube Goldberg device (Reed 1977). It has a keyboard in front that resembles, but is larger than, a typewriter, and around the back of the machine is a heating pot containing molten lead. When an operator presses the keys, he releases brass molds stored at the top of the machine. Each mold contains one letter, number, or symbol. As they drop down, they are assembled automatically in a line, in a metal holder near the keyboard.

The operator then transfers this line of molds, also called matrices, to a casting unit in the machine. The matrices line up in front of a steel cavity. Molten lead is squirted into the cavity mold and comes in contact with the brass matrices which contain the assembled letter images for the desired line. The lead hardens and retains the image of the brass-mold letters on its top surface. The cooled lead is then ejected from the machine as a precision-cast lead slug, known as a line of type.

The brass molds are automatically returned to storage at the top of the typesetting machine, ready to be released for use in succeeding lines. The entire operation has aptly been termed the "hot-lead" or "hot-metal" process.

The metal linotype slugs are then passed on to a handman who stands at a long, high table and proceeds to "make up" one or several pages from these slugs. While the page is being made up, or while the type is awaiting editor's orders, the proof is sent to a proofreader who compares it to the original copy and corrects it as necessary. When the page is completely made up and corrected, the "makeup" man locks it into place in a steel frame, called a chase. He signals a composing room assistant, who places the chase on a wheeled table and transports the page of type to the stereotyper.[1] This craftsman makes a cardboard matrix of the entire page, which is later transformed into a curved lead plate for letterpress printing.[2] The role of the machinist, who is not technically a printer but who falls under the jurisdiction of the printers' union, is to maintain the linotype machine in good working condition.

By 1976 with the exception of the *New York Post*, some news copy and most advertising copy at the *Daily News* and *The New York Times* were being produced with the help of a teletypesetter (TTS). The job of the linotype operator who is working on a teletypesetter-equipped machine has changed. He no longer punches the keys of his linotype machine to produce and set the type. Instead, he receives his copy in the form of a punched tape, which he threads onto his machine. After he punches a few coded instructions, his job is to "monitor" the machine as it automatically produces the slugs to be passed on to the handman for page makeup.

The production of the punched tape itself is now computerized. Typed and edited copy is passed from the newsroom to an operator sitting at a video display terminal (VDT). As the operator types a word on his specially equipped typewriter keyboard, that word will appear on the TV screen attached to his machine. The operator also inserts coded commands to the computer for spacing and sizing and width setting of the material.

The electronic impulses generated by the operator in the process of striking the keys are stored in the memory bank of the VDT. The operator then passes the story from the VDT memory into the main storage-and-processing computer, which automatically hyphenates and justifies (aligns the margins of) the story and produces both a high-speed typewriter printout for story identification and a roll of punched tape six holes wide, for use by the operator who is running a teletypesetter-equipped linotype machine.

The savings generated by the use of punched tape are considerable in both labor and time. For example, three automatic linecasters (as these specially equipped linotype machines are called), tended by one operator, can produce as much metal type as seven or eight linotypists (Kelber and Schlesinger 1967, p. 99). Moreover, many ready-to-thread tapes of news stories are obtained directly from the offices of the Associated or United Press, thus bypassing the composing room VDT operator altogether.

Now even the teletype process is on the way to obsolescence as composing rooms are switching from hot metal to what is called "cold type." The 1974 contract between Local 6 and the publishers of the three major New York dailies has given the publishers a green light to introduce any and all automated equipment. The publishers estimate that by the end of the 1970s cold type will have largely replaced hot metal and the workforce requirements for the composing room will be reduced.[3]

Cold type—defined as any printing process that does not involve hot metal—has been in use for almost thirty years, but only in the last several years has just about all advertising matter at the three dailies been set by one of the cold-type methods. The basis of the cold-type method is "photocomposition," that is, a process by which machines "set" type on film or photosensitized paper, instead of casting metal slugs. The process is continually being improved, according to one informant, who said that *The New York Times* has now seen four generations of photocomposition machines come and go. The most recent pride of management at *The New York Times* is the Metroset, called by those who like it the "miracle machine" and by those who do not a "technological Frankenstein." How does this showcase of the composing room work in the case of advertising matter, for example?

At *The New York Times*, a computer system with VDT terminals, known as the Harris 2200 system, combines the previous functions of one operator and two handmen into one VDT machine. First an operator types the advertising copy onto a paper, using an IBM Selectric typewriter equipped with an optical character-reading face of type. He pays no attention to size of type or width or position in the layout; he just types the copy "straight." This typewritten paper is then "read" by an optical character-reading machine, which can translate the typewritten copy into punched tape similar to the TTS variety.

This type, containing the "raw" advertising-copy words, is then given to the Harris 2200 VDT operator who runs it into his machine and has the "raw" copy come up on his VDT screen for further processing. The 2200 operator codes each line of the advertising copy according to size, width, and layout position, and then he produces a final punched tape containing all the information needed to produce the advertisement. The final Harris eight-level tape is then placed into the Metroset for actual phototypesetting.

The Metroset machine is a minicomputer processor about the size of a large bedroom dresser. In its computer are stored the digitized codes for thirty typefaces to match the Harris 2200 codes. When the Harris tape activates the Metroset, the correct typeface, size, and width appear on a miniature, high-quality cathode-ray tube within the machine. These digitized typeface images are exposed against a moving piece of photosensitized paper, at the rate of up to 1,000 newspaper lines per minute.

The exposed paper is developed and dried within three minutes in one of the composing room's Logeflow machines. The phototype material which emerges is then ready for the handman who will "paste up" the material (referred to by many of the men as "paper dolls") and assemble the parts of the advertisement as it will appear on the printed page.[4]

Equipment such as the Metroset eliminates all aspects of the hot-metal process and radically changes the nature of the work of the four categories of printers—operators, handmen, proofreaders, and machinists. Operators are no longer "kissing the keys" of the old linotype machine, but are either monitoring several automatic linecasters or sitting at IBM Selectric typewriters or video display terminals and preparing copy for automatic, electronic typesetting.

Handmen are no longer cutting the metal slugs into appropriate sizes for making up a page, but rather are receiving photographic copy of news and advertising matter and pasting these on sheets in preparation of the page. Proofreaders, in addition to reading proofs of hot type, are also receiving computer printouts of stories and advertisements containing both letters and computer codes, which they verify and return to the appropriate VDT operators for correction and processing. And machinists, many of whom have been retrained as electronic technicians, are now servicing expensive, complex equipment, rather than the old linotype machines.

Printers' Satisfaction with Their Work

Clearly, sweeping technological change in the composing room has dramatically transformed the work of the printer. Can we expect, however, that these changes have also altered the *meaning* of work for this group of skilled craftsmen? Blauner (1964), Friedmann (1961), and Woodward (1958), among others, tell us that technological factors impinge on just about every aspect of job satisfaction. And the literature on automation abounds with material describing the frustration of workers in automated shops and the fears of workers in shops about to introduce automation. Much of the frustration stems from boredom, monotony, routine, and dissociation from the final product as the skill of the craftsman is replaced by the machine. Much of the fear stems from concern, not only about the obsolescence of a skill, but about job security as automation displaces workers in increasing numbers.

Few of the printers worry that the security of their jobs is threatened by the introduction of automation into the composing room, for barring the closing of their newspaper, all printers at the three papers have been guaranteed lifetime job security, as well as built-in wage and cost-of-living increases over the eleven-year-life of the contract. This ironclad guarantee provides a unique opportunity to examine the response of those workers who have been exposed to the new processes, as well as the perceptions of those who have not, in a relatively pristine setting (that is, with confidence that what they say to an interviewer is not colored by fear of losing the job).

Past research has consistently shown high rates of job satisfaction among printers (Lipset, Trow, and Coleman 1956; Blauner 1960, 1964; *Work in America* 1973), and this research does as well.

Despite the changes that have taken place in the composing room, printers still are a very satisfied group. Almost three out of five find their job really interesting, one-third feel that it is all right, and only 8 percent say that their work is pretty dull and monotonous. (See appendix B for distribution of printers' responses to the survey items.) Similarly, almost three out of four agree that when they leave the composing room, it is with a feeling of a job well done. Some have mixed feelings about this, but only a handful disagree. There is some discontent perhaps among the two out of five printers who indicate that the work they do is too simple for them to use their ability. However, more than one-third think their work is indeed a challenge to their ability.[5]

Some twenty-five years ago Lipset, Trow, and Coleman conducted a study of printers in New York City's Local 6 and inquired into several aspects of printers' job satisfaction, thus permitting us to see if attitudes have changed since that time. Printers today are less likely than those of twenty-five years ago to feel that the prestige that printing receives from people outside the trade is excellent.[6] This change may reflect the fact that printers have lost the educational edge which they held over other blue-collar workers. Although they still deal with the printed word, printers are no longer the educated "elite" of the working class.[7]

The drop in the prestige with which printers view their trade may also reflect some of the changes that have occurred in printing during the past two decades as automation has been introduced into the composing room. As more than one printer noted, "Now it's work that any high school girl can do."[8]

The monetary rewards of printing, on the other hand, have clearly improved since the earlier study of the printers' union (Lipset, Trow, and Coleman 1956), and these improvements are reflected in the printers' ratings of their pay and job security. At that time, one-fourth of the printers rated their pay as only fair or poor, and almost the same proportion doubted that they had any job security. These days, only one in ten rates either his pay or his job security as fair or poor. Reflecting the terms of the new contract, which put the average earnings of the printer in 1976 well above $15,000 a year, more than one out of three rated the pay as excellent. Again, reflecting the lifetime security clause of the new contract, over one-half agree that they enjoy excellent job security.

The new contract provides all printers who were employed at the time of its signing with lifetime job security. Those who were on the substitute list at that time were also taken on as full-time journeymen with the same guarantee.[9] This guarantee extends beyond the life of the eleven-year contract (New York Typographical Union Number 6 1974):

> Regular situation holders and substitutes . . . shall hereafter be designated Guaranteed Employees. . . . A Guaranteed Employee shall not be subject to a decrease in the force. This guarantee survives the term of this Special Agreement and the contract . . . [and] shall enure to the successors and/or assigns of any Publisher signatory to this Special Agreement as if such successor or assign had been an original signatory thereto. Only in the event of a permanent suspension of the Publisher's newspaper will the employment guarantees of this Special Agreement cease.

Unless the newspaper closes down, all printers at the three papers can be certain that their employment is secure.[10] Accordingly, more than nine out of ten rate job security favorably. There are a few skeptics who ask, As far as lifetime guarantees—is there a lifetime guarantee on anything? Or, as the printer from the *New York Post* commented, "Every night we pray for the "old lady's" [the publisher's] health. Life-time guarantee or not, our job depends upon her."[11]

Younger men who will still be around when the contract expires are somewhat less sanguine about job security than the older printers. Fewer than half of the younger printers, compared with three out of five of those who plan to retire within the life of the contract, rate their job security as excellent.

Human relations experts emphasize the importance to job satisfaction of a compatible social atmosphere at work. Indeed, Lipset, Trow, and Coleman reported in 1956 that many of the printers socialized off the job extensively with other printers, typically with those in the same shop where they worked. While this is not the case today (only a few see fellow workers off the job), printers do like their coworkers; an overwhelming majority agree that most of people in the shop are nice to work with.

One of the most reliable indicators of job satisfaction over the last few decades is the question What type of work would you try to get into if you could start all over again? (Blauner 1960; *Work in America* 1973). In the early 1950s almost two-thirds of the printers said that there was no job other than the one they now held which they would like to go into, and close to one-half said that if they were starting all over again, they would choose printing.[12]

At the time of our study, only one-third said that they would choose printing again. It is unlikely, however, that this low figure is a reliable indicator of printers' satisfaction with their jobs today. Rather, it reflects the recognition that opportunities in printing are shrinking rapidly and that in New York City, at least, the door is closed to new printers. Many who opt for another occupation echo the sentiments of the handman who said, "I love printing, but it's a dying field. There's more opportunity for a young man in other occupations." or of another who said, "There will be no printing in a matter of years because automation is eliminating jobs."

Variation in Job Satisfaction

Some printers, of course, are more satisfied than others with the various facets of their jobs, but few factors differentiate the more satisfied from the less satisfied printer. Everyone in the composing room earns approximately the same amount of money, so it is hardly surprising that they are alike in how they rate their pay.[13] The older men are somewhat more satisfied than the younger ones with their work, particularly with the security of the job and the compatibility of their fellow workers. Overall, however, regardless of age, printers agree and are positive about their work for the most part. Similarly, satisfaction levels are consistently high for printers, regardless of the number of years of schooling they may have had.

Even a printer's job classification (operator, handman, proofreader, or machinist) makes little difference in how he views his job. Operators and handmen, who comprise more than three-fourths of the composing room force, are in substantial agreement (and quite positive) about the interest and satisfaction they get from their job. In 1976 roughly one-half of the printers, however, had been retrained and were working with the new automated equipment while the other one-half still used the old, familiar techniques. The apparent consensus between operators and handmen about job satisfaction may be masking differences between those who work with the new technology and those who do not.

Appealing though this explanation may be, the facts do not bear it out. Retrained or not, operators and handmen still generally agree about the interest and satisfaction inherent in their jobs.

That the bulk of the composing room force—retrained or not—maintains a high level of job satisfaction runs counter to all expectations. Giant change has occurred, but even though not every printer can be described as glowingly satisfied with his work, the overall level of satisfaction in the composing room is high. Can it be that changes in technology have little or nothing to do with printers' attitudes toward their jobs?

One hint that changes in the nature of the printer's work have affected his view of his job comes from looking at the proofreaders and machinists. These two groups comprise only a very small proportion of the workforce,[14] but their attitudes toward their job present a study in contrasts. Machinists are much more likely than proofreaders to feel that the job is really interesting and to report that when they leave the composing room at the end of the shift, it is with the feeling of a job well done. Similarly, almost no machinists, but most of the proofreaders, complain that the work they do is too simple for their ability.[15]

The contrast between proofreaders and machinists is even sharper if we separate those who have been retrained from those who have not. Among those who are still using the old methods, proofreaders report somewhat higher levels of job interest and fulfillment than experienced by machinists. But among those who have been retrained and are working under the new system, the machinists' satisfaction skyrockets while that of the proofreader plummets. This difference between the retrained proofreaders and machinists holds whether they are talking about the interest of the job, the challenge to their ability, or the satisfied feeling derived from a job well done.

Proofreaders and machinists represent only a small proportion of the composing room force, and separating them further into those who have and those who have not been retrained yields numbers much too small from which to generalize. Still, the strong and consistent differences between them, as well as the absence of difference between operators and handmen, point to the need for a very careful examination of the specific changes in the nature and organization of work which result from the automation of a complex, skilled craft.

A Technological Approach

The link between technological change and job satisfaction has been posited in a fairly large and respectable literature.[16] Technology, it is argued, imposes limits and constraints on interaction among fellow workers, on supervisory relations, and on how much control the worker can exercise over the pace, quality, and scope of his tasks—in sum, over all those conditions which research in the sociology of work has found to affect job satisfaction. Most generally agree that the automation of a craft results in lower job satisfaction and more alienation because such automation is likely to reduce worker control, initiative, autonomy, freedom of movement, or interaction with fellow workers.[17]

This school of thought constitutes the starting point for our analysis of the response of printers to the changes which have taken place in the composing room. We draw on their discussion to speculate about the way in which printers might be expected to respond to the specific technological changes that comprise the automation of their craft. Then we propose an alternative approach for understanding the printers' response to automation.

Control over the Immediate Work Process

Blauner (1964) has argued that the high level of job satisfaction traditionally expressed by printers stemmed from the fact that, unlike most blue-collar workers, printers have always exercised considerable control over the immediate work process. "A printing craftsman not only sets his own work rhythm, he is free from pressure on the job" (Blauner 1964, p. 42). Similarly, Lipset, Trow, and Coleman (1956, p. 23) have noted that while printers do not control the overall policies of the enterprise, the "nature of the job control exercised by the International Typographical Union (ITU) has meant that to a considerable extent the workers run the composing room."

The new technology has changed the *content* of the printer's job, but our months of observation in the three composing rooms uncovered few changes in the factors which advocates of the technological approach consider critical to job satisfaction. Except under pressure of getting out the first edition, printers still control the rhythm of their work. They can speed up, slow down, or take a short break at their own discretion. The pressure in the composing room mounts as the deadline for the first edition nears, but this has always been the case and holds for those operating both the old and the new equipment. Moreover, many of the men working on the "edition shift" (during which the first edition is rushed to press) enjoy the challenge of this daily race against the clock and echo the sentiments of one operator who, when asked why he prefers the night (edition) shift, said, "The edition shift is the most exciting—you actually put the paper to bed."

Freedom of Physical Movement

Another ingredient of worker satisfaction is freedom of physical movement—the ability to step away from the machine for a few minutes, to walk around the shop. Printers have traditionally enjoyed this freedom. In 1947 a Roper survey for *Fortune* magazine asked, "Is yours the kind of job on which someone would have to take your place if you had to leave your work for a half hour or so, or could you let your work go for a half hour and catch up on it later?" Four-fifths of the printers, compared with two-thirds of the entire sample, said that they could leave their work for a half-hour with no problem (cited in Blauner 1964).

We did not ask this question, but it was quite apparent that at all three newspapers the printers, whether working with the old or the new equipment, still enjoy freedom of physical movement. They visit with one another, talk about a work assignment, read the bulletin board, go to the coffee machine, or walk into the office of the chapel chairman with a question or a complaint.[18] As one operator remarked, "You can take time off for coffee and no one puts pressure on you if you take a couple of extra minutes. As long as you do your job—they don't care."

On the other hand, some of the men operating the VDTs or Harris 2200s report that this work is physically *more* demanding than their old jobs because "You have to sit at a typewriter in one spot; I'm not used to that and it makes me tired." Even these men, however, are not tied to a conveyer belt; they can and do get up and stretch or take a break at their discretion.

Variety, Purpose, Meaning

A major component of job satisfaction, according to the technological school, is the extent to which a person's work affords variety, purpose, meaning, or challenge (Friedmann, 1961, pp. 65-66).[19] The work of the printer has traditionally provided the opportunity for these ingredients. Long years of training have equipped the printer for almost any kind of work necessary to prepare the finished page. While there is some degree of specialization and division of labor, work in the composing room does not involve the elaborate subdivision of tasks characteristic of assembly-line technology. The majority of men have had experience with all phases of the work. As many of the men said with pride, "There's no phase of this operation that I can't handle. They can put me on anything, and I'll turn out a good job."

Even within classification boundaries, the range of tasks may vary. While some operators may have been at the linotype machine for most of their working lives, the majority have spent time on a range of equipment—the relatively simple monotype machine, the standard linotype, the teletypesetter, the IBM Selectric typewriter, and perhaps the more complex automated keyboard processes. The handman may also move from making up advertisements to marking up pages or cutting and pasting up "paper dolls." Similarly, the machinist is responsible for the maintenance of the full range of equipment in the composing room. And while proofreaders may remain in the proofroom for many years, each day the material they handle is different. As one proofreader said, "There's no humdrumness about it. Every night is a different ball game—a brand new page of history." An operator, in the trade for forty-six years, added, "It's exciting, not at all boring—never the same routine every day."

The new technology has not removed this element of variety from the work of the printer. In fact, the 1974 contract gave management discretion to move men from one task to another in the shop, a right until then within the province of the chapel officers. Some men resent this new prerogative, largely because they feel that it cuts into the traditional control of the union over the composing room. Others, however, welcome the opportunity for increased variety in the workday.

Identification with a Final Product

The ability to identify with a final product, even though one may be working on only one segment or phase, is seen as another major source of job satisfaction. Despite automation, the comments of the printers suggest that many have such a sense of identification and agree with the operator who said, "It's like when an artist sees a painting he's finished. You see someone on the subway reading an article that you've set up, and it's a feeling of accomplishment." Or, as one handman noted, "It's a good feeling to set up an ad. When you see it in the paper, you can tell your wife you worked on it." Unlike the automotive worker who can never be certain that he has placed the rear-view mirror on the car in his neighbor's driveway, the printer knows full well that he has set the type, pasted up the advertisement, or proofed the material in the newspaper that his neighbor is reading.

Pride of Craft

Perhaps the ultimate concern of the technologists is whether automation cuts into (if it does not eliminate completely) the craftsman's pride in his long-held skills. Such pride, it is argued, is a critical ingredient of job satisfaction; and lack of such pride, a harbinger of alienation. Kelber and Schlesinger (1967, p. 69), commenting on the introduction of photocomposition in the late 1950s, note that the new techniques ". . . did not eliminate, but rather elevated, the craftsmanship involved in mixing typefaces, marking up copy, setting type in unusual configurations, and meeting the requirements of the most complicated layout."

For some printers, however, pride of craft has diminished. Sometimes it was explicit, sometimes only a shrug of the shoulders, but all the same, we could sense the fear that automation was rapidly making valued skills obsolete. As one printer said, "There's much less room for personal expression. Before, I took pride in my craftsmanship. It was an accomplishment working in metal. Now the skill is gone." Another printer commented, "The automated processes take away the beauty of the printed page. For a *real* typographer there's no pride left—printing as a craft is finished." A seventy-year-old operator, in the trade for forty-eight years, summed it up: "Printing has lost its zing!"

Typically, such comments came from the printers sitting in classroom style at Selectric typewriters (after passing a forty-word-a-minute typing test). Almost to a man they looked bored, uncomfortable, or at best resigned. When we asked one older printer, who looked particularly unhappy as he sat at the typewriter, how long he had been working at the Selectric, he sighed and said, "Three very long months." To some extent, then, the new technology appears for some workers to have been accompanied by a loss of pride and a diminished sense of craftsmanship.

For many men, however, this was not the case. Some were clearly delighted with their new work and agreed with Kelber and Schlesinger (1967): "The equipment is very sophisticated. It takes a lot of skill, and I think I'm pretty good at it." Another printer in the trade over thirty years and now working at a VDT said, "I can see what I'm creating, and it gives me a real feeling of self-satisfaction. Putting ads together this way is a creative skill—it's a whole new craft to learn."

Twenty years ago, William Faunce (1958) studied the adjustment problems of 125 automobile workers in four of the most highly automated departments of a large plant. He found that increased feelings of pride and satisfaction "were related for the most part to the complexity and cost of the machine tended by the worker." In the same way, for some printers, pride in their own skills has been transferred to the new complex electronic equipment. The men at one newspaper seemed to be fascinated with the "miracle machine." They took an almost personal pride in demonstrating the speed with which a piece of copy could be prepared for paste-up. Printers working at the VDTs appeared to take similar pride in the processes involved in sizing, spacing, and aligning an advertisement and then storing the result in the computer's memory. Indicative of such feeling was the frequency with which we were stopped as we walked through the composing room so that a newly trained operator could demonstrate the capabilities of his equipment.

Supervisory Control

Finally, a crucial aspect of the job which has been found to impinge on job satisfaction is the degree of supervisory control exercised over a worker's performance.[20] Skilled craftsmen, who have internalized standards of workmanship, are particularly likely to resent such control from foremen whose level of competence may hardly differ from their own. Our concern, however, is not whether the printers are satisfied with their supervisors, but rather whether the new technology has altered the worker-supervisor relationship. None of the supervisors with whom we spoke felt that automation had in any way changed the nature or degree of their control over their subordinates. The printers themselves tended to concur. Some disparagingly referred to the supervisors as "bow-wows," because "they bark at us all the time" and agreed with the printer who said, "There are too many 'bow-wows' looking over your shoulder all the time. I don't like anyone telling me what to do, especially when they don't know as much about it as I do."

In contrast to these complaints about their supervisors, many printers indicated that one of the major assets of the job was the *lack* of pressure imposed by direct-line supervisors or higher-level foremen: "You don't feel that you're always being watched. They give you a lot of freedom, and that's important.'' Or, ''I like the fact that I'm left alone. No one's pushing or bothering me. If you do your job, they don't bother you—you can work at your own pace, and no one's on your back.'' With only a few exceptions, such as one printer who wished his supervisors would be more helpful when problems arose, these craftsmen define good supervision as less supervision. They like their jobs precisely because they can work at their own pace; no one pushes them or looks over their shoulders.

Whether the printer expressed pleasure or displeasure with the nature of supervisory control, no one suggested that the situation was any different depending on whether he was working with the old or the new equipment.[21] Apparently the relationship between printers and their supervisors has not changed as a result of the new technology—supervision is neither looser nor tighter than previously. We have little reason, therefore, to expect the printers' response to automation to be colored by changes in the nature or extent of supervisory control.

While the nature of supervisory control over the immediate work process has not changed with automation, a significant change has occurred in the degree of external control exercised over the printer. Previous to the new contract, ''The job . . . [belonged] to the man, rather than to the foreman or the shop'' (Lipset, Trow, and Coleman 1956, p. 24). A printer who decided to take off a day to go fishing, or a week to go to Florida, could hire a substitute to take his place. No longer is this possible, for the substitute system ended with the new contract. Similarly, in the past for a man to be moved from one part of the composing room to another required union consent, and a printer with high priority (seniority) enjoyed first choice when new assignments were made.[22]

With the new contract, this is no longer the case. The introduction of automated equipment has led to a variety of new tasks that a printer may be trained to perform, and management has complete discretion to move men around the composing room—without regard for priority. Some printers welcome this opportunity for increased versatility and variety in the day's work. Others, however, resent the erosion of their traditional control over job conditions in the composing room. Although members of the union themselves, supervisors are the representatives of management on the composing room floor, and in this capacity they will increasingly be responsible for shifting men from one operation to another. Accordingly, as the pace of automation accelerates, supervisors—as the instrument through which printers' long-cherished control over job conditions in the composing room is being diminished—may increasingly become the target of the printers' resentment.

For almost a century the technology of printing, combined with the strength of the ITU, has provided union printers with relative freedom to control the conditions of their work, to set their own pace, to minimize supervisory constraints, and to maximize a sense of pride of craftsmanship. As a result, compared with most blue-collar workers, printers have enjoyed high rates of job satisfaction and have experienced little of the feelings of powerlessness and meaninglessness which are components of worker alienation. The tasks which the printer performs are different today with the advent of automation, but those aspects of the old technology which may have facilitated the relatively advantageous working conditions of the printer have not radically changed as a result of the new technology. For those performing certain operations, work is perhaps less meaningful; but for most of those who are being retrained, there are new skills to learn, complex equipment to operate or service, the same daily challenge to meet deadlines, and the same relative freedom from close supervisory control. Management's new discretion to shift workers from one task to another is a source of resentment for some printers, but is welcomed by others. Clearly, the response of the printers to automation cannot be predicted solely on the basis of the nature of the technological changes taking place in the composing room.

Orientations toward Work

The effects of these technological changes must be supplemented by an understanding of the orientations of these blue-collar workers which mediate between the technological conditions of work and the response to automation.

One school of thought, stemming largely from British industrial sociology, disputes the contention that technological considerations are necessarily causal factors in worker satisfaction or alienation. The studies of Goldthorpe, Lockwood, Beynon and Blackburn, and others in England suggest that workers bring with them to the workplace different orientations toward the meaning of work and that these orientations mediate between the technological features of the work situation and the nature of the response (Goldthorpe et al. 1968). As Beynon and Blackburn (1972, p. 3) remark, "The extent of the influence of technology upon the ideas of workers . . . cannot be directly deduced from the physical fact of the machinery, but from the interpretation of the machinery by the actors in the work situation." Such interpretations derive from the orientations either which workers bring to the workplace or which evolve from the sociotechnical conditions of work.

Most blue-collar workers today, according to this school of thought, are instrumentally oriented in that their paramount concern is with the extrinsic rewards of the job—wages, fringe benefits, security of employment. In other words, they view their job primarily as a means to the end of maximum economic return that will permit the good life (Goldthorpe et al. 1968, pp. 36-38). The dichotomy between the work world and the nonwork world is a sharp one; instrumentally oriented workers rarely extend their relations with fellow workers beyond the gates of the plant, satisfactory as these relations may be within the enterprise.[23]

Similarly, instrumentally oriented workers tend to see their union primarily as the vehicle for maximizing the extrinsic rewards of work. While commitment to the union may be strong, it is a pragmatic, rather than ideological or fraternal, commitment (Beynon and Blackburn 1972, p. 178). In the same way, they tend to view their bond to the firm in economic, rather than human relations, terms; they see this bond as a symbiotic relationship in which they provide their labor in return for money and security (Goldthorpe et al. 1968, chapter 4).

The technological conditions of work might be such as to lead to a prediction of high levels of dissatisfaction or frustration (Friedmann 1961; Blauner 1964), and workers may indeed voice discontent with the degree of interest, challenge, or variety of their work. But if greater priority is attached to the extrinsic rewards of work and workers are content with these rewards, then overall satisfaction with the job is likely to be high.[24]

Although, according to Goldthorpe et al., today's blue-collar worker more often than not is instrumentally oriented, many workers maintain what these authors have called a "traditional" orientation toward work, an orientation in which one's labor is seen as not only an economic activity, but also a group one, constituting a central life interest. The worker with a strong component of traditionalism tends to be a skilled craftsman for whom work means far more than economic return. Pride in workmanship, internalized craft norms, high standards of workmanship derived from long years of training and apprenticeship and reinforced by close-knit relationships with fellow craftsmen—all these engender a feeling of solidarity and a special identification with work.[25]

The strong ties with coworkers characteristic of traditionalists often take the form of what has been called an "occupational community." Such communities, made up largely of workers in the same trade, are likely to develop in occupations that are isolated either spatially (miners or lumbermen) or by peculiar hours of work (firemen, steelworkers, printers). The fundamental characteristic of the occupational community is that workers tend to socialize in their off-hours more with people in their own line of work than with people in other occupations. These informal social networks, in turn, tend to reinforce the traditional orientation of the skilled craftsman and strengthen the centrality and group nature of work.

Printing is perhaps the classic example of a trade in which occupational communities became a hallmark. Lipset, Trow, and Coleman, describing this strong occupational substructure of printers in New York City, found that contact with coworkers did not end with the shift. On the contrary, printers bowled together; they had a drink together after work; at least one of their best friends was, more often than not, another printer.[26]

A number of factors combine to explain the development and persistence of occupational communities in the printing trade.[27] Dealing with the printed word, as these craftsmen do, calls for a degree of literacy which, until recently, set printers, in their own eyes, a cut above the average blue-collar worker.[28] They developed a sense of occupational pride and prestige which made them turn to one another for company and thus isolated them, socially if not spatially, from other manual workers. Reinforcing this social isolation was the fact that more printers than not worked on a night or "lobster" (after-midnight) shift,[29] making it difficult to socialize with the vast majority of persons in other trades.

Other factors endemic to printing increased the possibilities for social contact and interaction among printers. Relative freedom from time pressure, light supervisory control, and ease of movement about the composing room all facilitated opportunities to socialize with coworkers on the job and laid the groundwork for friendships which continued off the job.

Further, the "substitute system" facilitated social contacts among printers. Those without work for the day would gather before the start of a shift, hoping for an assignment. If none was forthcoming, rather than go home and then return to try their luck for the next shift, groups of substitutes would pass the hours between shifts bowling, eating, or seeing a movie. And finally, according to union rules, a journeyman who was ill or planning a vacation would select his own replacement from among the substitutes; accordingly, it made sound economic sense for substitutes to cultivate friendships with working printers.

All these factors facilitated the development and supported the growth of what Blauner (1964) called a "highly developed off-the-job occupational community." Blauner (1964, p. 50) noted:

> The printer's leisure life is full of voluntary associations made up of fellow printers: social clubs, fraternal orders, ethnic lodges, veterans' groups, technical clubs, discussion groups, bowling societies, and other sports teams. These associations provide an organized independent group life which is unusual for factory workers in modern American industry.

Since an occupational community tends to reinforce a traditional orientation to work, and since the occupational community of printers has been well documented, we expected that the response of the printers to the automation of their craft might be negative, expressing resentment over obsolescence of skills. It hardly seemed possible that this group of older craftsmen, with their strong informal social networks and keen sense of the prestige of their occupation, would view the changes in the composing room with equanimity.

At question, however, is whether the occupational community of printers is still the viable phenomenon described by Lipset, Trow, and Coleman. Much has changed in printing and, in particular, among New York City's printers. Printers who are out of work still gather at union headquarters in hopes of finding employment, but regularly employed printers can no longer appoint their own replacements from the "slipboard" (substitute list) should they be ill or wish to take off a few days. The end of the old substitute system removed one of the major underpinnings of the printers' occupational community.

Further, many printers may still harbor a sense of pride and prestige because they are dealing with the printed word, but today these workers have little educational edge over most other blue-collar workers. There is less reason, accordingly, for them to shun the company of workers in other occupations and to band together for their off-the-job activities.

Another factor which sociologists claim provides a basis for occupational communities is "unusual" hours. True, printers still work on one of the three shifts, but today working unusual hours does not facilitate off-the-job interaction with coworkers to the same degree that it might have twenty-five years ago. Most printers on the night or lobster shifts told us that they spend their leisure time watching television, working in the garden, doing home repairs, or just relaxing. No longer do printers socialize extensively with one another. Rather, as one foreman remarked, "They scatter like quail at quitting time."

While there may be specific factors endemic to printing which have contributed to the breakdown of the occupational community, a number of external and more general pressures are expediting the disappearance of such communities—whether in the printing trade or in other occupations. In fact, the demise of the occupational community of printers is probably just one particular instance of a more general phenomenon. Workers in the United States have become part of what social commentators call postindustrial society.[30]

The average worker today lives in the suburbs—fully 60 percent of the printers in our sample own their own home and commute to the city from Long Island, New Jersey, Staten Island, or Connecticut each day. Their off-the-job time is spent largely at home with their families. Consumerism has replaced the formerly salient place of work in a man's life (Alt 1976). Satisfactions may be derived from the job itself, but even more important are the consumer goods that one can acquire as a result of the job. Status, too, derives less from what a man does than from what he has to show for what he does—two cars, children in college, a vacation trip. Organizational attachments, if they exist, are generally to large, formal, bureaucratic associations in which personal involvement is minimal. Informal contacts are generally restricted to backyard barbecues with one's family or neighbors, and one's neighbors are hardly likely to overlap with fellow workers.

Occupational communities—whether in the classical sense of people in the same trade living together in isolation or in the sense of off-the-job, informal social contacts with fellow workers—are rapidly disappearing in this consumer-oriented mass society. Printers are an important example because one can see clearly that the pressures contributing to the breakdown of the printers' occupational community are internal as well as external. For many years, technological factors and the general organization of work in printing insulated the printers to some extent from the general trends in U.S. society which have contributed to the disappearance of occupational communities. But in recent years, and particularly since the 1974 contract, these endemic factors no longer exist. As a result, the occupational community of printers may well be a phenomenon of the past, lingering perhaps in the memory of the older printer but hardly a reality for most.[31] Repeatedly, when we asked the men whether there was much off-the-job socializing, they said, ". . . there used to be a lot of that, but not anymore. Years ago, you would go in after work and have a couple of beers with the fellows; not today."

The disappearance of the occupational community of printers may signal a weakening traditional orientation in favor of a more instrumental one. Although work may be important, home, family, security, the ability to afford the good things in life—these may be more important to today's newspaper printer, and if these are satisfactory, perhaps changes in the composing room will be accepted with relative ease.

Chapter 2 examines the response of the printer to the technological changes which have occurred at the workplace. We consider two groups of printers: those who have been retrained and are already working with the new equipment and those who have yet to be retrained. This provides an opportunity to compare the actual experience of one group with the expectations of the other. Then in chapter 3 we see how the orientation of the printer to his work mediates between technological change in the composing room and his response to this change.

Notes

1. To observe this entire process in operation during the half hour before the first edition was a thrill which did not abate even after a month or two in the field. On a large board red numbers were electronically displayed, representing each page of the newspaper. As the makeup man

completed a page, locked it into place, and sounded the call for the carter, that number on the big board was eliminated. One could feel the excitement and tension mounting as the deadline neared and a score of red numbers still remained on the board. One could equally detect the release of tension, and almost deafening quiet in the composing room, when the last page (usually page one) was completed and the board was empty.

2. Stereotypers do not fall under the jurisdiction of the ITU; they have their own union.

3. Management at *The New York Times* told us that by summer of 1978 they planned to have switched completely to cold type—linotype machines will no longer be seen in the composing room.

4. In the case of newsmatter, the process is essentially the same. The typed, edited, and proofed copy goes through the VDT process rather than the Harris 2200 machine, and the electronic impulses are stored in the VDT computer until activated by the Metroset machine.

5. Side by side with those who are operating or servicing the old linotype machines, or setting type by hand, are those who are now working with the new automated equipment. Later we will differentiate between these two groups; the results cited here are for the sample of newspaper printers as a whole—retrained or not.

6. We suspect that some of the prestige which printers attribute to the trade is actually "borrowed prestige" from the newspaper at which they work, for 27 percent of the printers at *The New York Times*, compared with 21 percent at the *Daily News* and 19 percent at the *New York Post*, rated the prestige of the printing trade "excellent."

7. In 1950 the median number of years of schooling of printers was 10.4 while that for all craftsmen, foremen, and kindred workers was 9.5 (U.S. Bureau of the Census 1953). Twenty years later the gap between the two groups had narrowed to 12.3 for printers and 12.0 for all craftsmen and kindred workers (U.S. Bureau of the Census 1973).

8. This remark hints of sexism, but we do not think that it was meant as such. Printers were telling us, in essence, that formerly their craft required long hard years of apprenticeship. Today, in contrast, the new processes demand only an acquaintance with the basics of typing—"women's work" in the eyes of this older blue-collar group of men.

9. Substitutes were printers who had completed their apprenticeship but were still awaiting regular full-time employment. They were available to fill in for ill or vacationing journeymen.

10. Control of the *New York Post* did, in fact, change hands about two years after the contract had been in force, and the new publisher must abide by the terms of the contract.

11. Follow-up conversations with printers at the *New York Post* indicate that the men in the composing room, especially the younger ones, are considerably more relaxed about job security now that the paper has changed hands.

12. Data are from the files of the ITU study (Lipset, Trow, and Coleman).

13. The pay scales for night and lobster (after-midnight) shift workers are slightly higher than that for day men, and this is reflected in the ratings of this item.

14. See appendix B for the distribution of printers' responses by job classification.

15. Although machinists are more satisfied than the other categories of printers with the intrinsic aspects of their work, they are not enthusiastic about the prestige of their trade: Almost none feels that the prestige of the occupation is excellent. This jaundiced view may stem from the machinist's status as "low man on the totem pole" in the composing room. Time and again, both supervisors and journeymen told us, "They [the machinists] are not *really* printers; they just service the machines." Since automation, however, the status of machinists has been enhanced. Many have taken special courses which train them to be "technicians," and their new tasks involve servicing the expensive, complex equipment which is being introduced into the composing room. As a result, we find that of all the categories of printers, retrained machinists are the most enthusiastic about automation. Most feel that their trade enjoys considerable prestige, while this is the case for only two of the seven machinists still tending the old machines.

16. Some of the classics in the field are Woodward (1958), Friedmann (1961), and Blauner (1964). Among the more recent treatments are Beynon and Blackburn (1972), Sheppard and Herrick (1972), and Braverman (1974).

17. Blauner concludes, however, that worker alienation is at its highest among those on the conveyer belt or assembly line, rather than among workers in continuous-flow or automated production (Blauner 1964).

18. In printing parlance, each shop in which union printers work is known as a chapel. The chapel chairman is the elected leader of the shop.

19. Blauner (1964, p. 117) agrees but cautions, "The average worker is able to make an adjustment to a job which, from the standpoint of an intellectual, appears to be the epitome of tedium."

20. For reviews of the literature on this subject, see Blauner (1960), Kahn (1972), and Carroll (1973).

21. The survey data support this observation. Comparing those who were and those who were not working with the automated equipment, we found that 25 percent of the former and 21 percent of the latter were displeased with the nature of supervisory control.

22. Forty-five years ago Perlman and Taft (1935, p. 51) wrote that printers, through the ITU, have "the most complete control over job conditions of any union in the world."

23. Goldthorpe et al. (1968, pp. 54-55) suggest that the technological organization of work may facilitate or constrain the pattern and frequency of

interaction on the work floor. On the other hand, these are not significant determinants of the *quality* of work relationships but must be combined with the workers' orientation to explain the presence or absence of close-knit, affective work relationships.

24. Beynon and Blackburn illustrate this point with an example of night-shift workers in an English factory. These men expressed numerous complaints about the monotonous, uninteresting character of their tasks. However, as a group they were instrumentally oriented, and they attached relatively low importance to the need for intrinsically satisfying work. The salient rewards for this group were the economic ones, and with these they were generally satisfied. As a result, their overall job satisfaction was relatively high (Beynon and Blackburn 1972, p. 86).

25. It would be simplistic to suggest that a traditional orientation is the direct antithesis of an instrumental one. More realistically, we agree with Goldthorpe et al. (1964, p. 41) that "*all* work activity in industrial society at least, tends to have a basically instrumental component . . . " which combines with greater or lesser components of traditionalism. Goldthorpe et al. note further (p. 39) that there are other directions which deviations from an instrumental orientation might take, for example, a bureaucratic direction in which the primary meaning of work is "service to an organization . . . in return for a career."

26. Canon (1967) demonstrates the existence of a "powerful occupational community" among compositors from twenty-eight firms in the London area and suggests that one consequence of membership in the community was the persistence of leftist political norms among this group, which in most other respects resembled the politically more conservative middle class.

27. For evidence of informally organized occupational communities of printers, dating back as early as the sixteenth century, see Lipset, Trow, and Coleman (1956, chapter 2).

28. In fact, printers did have somewhat more schooling than the average manual worker (see note 7).

29. Several newspapers (all but one, now defunct) were located near the docks. The term *lobster*, so we were told, refers to the fact that the men on the after-midnight shift at these papers were having their "lunch" break just at the time that the fishermen were hauling in their day's catch of lobsters.

30. For several examples of the vast literature on this subject see Pizzorno (1967), Goldthorpe et al. (1969), Westergaard (1973), and Alt (1976).

31. Our study did not extend to the "book and job" (commercial) shops which are considerably smaller than the newspaper chapels. Lipset, Trow, and Coleman (1956) found a greater degree of camaraderie and off-the-job interaction among the men in these establishments.

Appendix 1A

Table 1A-1
Printers' Ratings of Prestige, Pay, and Job Security
(percent)

	Prestige		*Pay*		*Job Security*	
Rating	*1950*[a]	*1976*	*1950*	*1976*	*1950*	*1976*
Excellent	41	23	22	34	35	53
Good	47	51	53	57	43	38
Fair or poor	12	26	24	10	22	9
(*N*)	(493)	(403)	(500)	(406)	(492)	(406)

[a]From the files of the ITU study.

Table 1A-2
Printers' Ratings of Several Aspects of Their Work by Classification
(percent)

Percentage Who:	*Operators* ($N=147$)	*Handmen* ($N=165$)	*Proofreaders* ($N=34$)	*Machinists* ($N=17$)
Find the job really interesting	61	53	44	59
Leave with a feeling of a job well done	72	72	59	88
Do not find the work too simple for their ability	32	29	30	76
Rate the prestige of printing "excellent"	25	23	26	12

Table 1A-3
Percentage Who Find Their Job Really Interesting

Job Classification	*Retrained*	*Not Retrained*
Operators	60 (72)	63 (69)
Handmen	52 (112)	56 (47)
Proofreaders	31 (17)	56 (17)
Machinists	70 (10)	43 (7)

2 Experience and Expectation

For almost a century, Local 6 has maintained strict control through its priority (seniority) system over the assignment of printers to any new equipment in the composing room (Kelber and Schlesinger 1967). Now, for the first time, the new contract permits management to introduce any kind of automated equipment it likes at whatever pace it deems appropriate and to assign printers as it wishes to operate the new equipment.[1]

Who Has Been Retrained?

By spring of 1976, 60 percent of the men had had some experience working under automation, and 40 percent had not. Management has clearly exercised its prerogative to assign men to different kinds of work in the composing room regardless of priority, for printers with high or low priority are equally likely to have been trained in the new processes.

Although many volunteer, the selection of trainees is largely made by management according to production needs.[2] As an official at one newspaper explained,

> First we determine which groups are most needed, that is, whether we need men to perform keyboard or non-keyboard operations. It's more efficient to retrain a man who has been used to the keyboard [linotype operators] for the new keyboard processes. We also have to juggle according to shift requirements; men on the night [edition] shift can't be as easily spared for retraining. Of course, we try to find out who would *like* to be retrained and then pick from this group according to our needs and, finally, by the men's priority.

Whatever the reasons, trainees have not been selected randomly. Management has passed over older printers and those with less schooling. Fewer than half of the printers sixty years or older, but three out of four of those under forty-five, have been retrained. Similarly with education—just two out of five without high school diplomas, but more than three in four with some college have been selected. Age and education, of course, are highly correlated, but management's preference for retraining the younger printer is particularly clear when we examine the joint effects of age and education on experience with automation. A young printer *without* a high school diploma has had about the same opportunity (61 percent) to work with the automated equipment as an older one *with* some college (65 percent).

Men who aspire to a supervisory position—virtually the only channel for promotion among printers—are also likely to have been selected for retraining. Among those who would like a supervisory post, three-fourths have worked with the new equipment, compared with one-half of those with no such aspirations. Particularly among the younger men, the desire for promotion correlates highly with selection for retraining.[3]

In the 1960s when teletypsetters and early versions of photocomposition equipment were being introduced at some of the New York City newspapers, the union charged that the linotype operators were being given preference over handmen in learning the new techniques. Today this is not the case: handmen are more likely than operators to have had experience with the automated equipment, perhaps because virtually all advertising matter (in contrast with newsmatter) is now set by the newer cold-type method, and handmen are largely responsible for advertising matter.

Well over one-half of the printers in our sample have been retrained; that is, they have learned how to type (at least forty words per minute), mastered the technique of adjusting copy on the VDT, discovered how to operate the electronic "miracles of the composing room," and, in the case of the machinists, taken courses enabling them to service the range of new, complex machinery which together comprise the cold-type process.

The Response of the Retrained Printer

How has it gone for these men? We asked the printers whether, compared with the old processes, they have found working with the new equipment physically more difficult or easier, intellectually more or less challenging, and more or less interesting.[4] Overall, three out of five printers have found working with cold type physically less demanding than the hot-metal process, and about one-half feel that the new methods are intellectually more challenging as well as more interesting.

Although some printers have had several years of experience working with the teletypesetter and other automated equipment, for most of the men retraining has been quite recent. By spring of 1976 about two out of five of the retrained printers had been working with the new equipment for less than one year, the same proportion for between one and two years, while just one in five claimed three years or more experience with the new processes.

How long printers have been working with automation, however, has little to do with their attitudes about the new equipment: the experience does not seem to pall as times goes by. Those who have worked at the new tasks for less than six months are no more likely than those who have had three or more years' experience to find that, compared with the old ways, the new equipment makes their work more interesting and challenging.

Nor does the amount of schooling a printer has had affect his response to working under automation. Men with some college and men with only a grade school education are equally likely to find the new methods intellectually more challenging and interesting.

Further, and contrary to our expectations, the age of the printer makes little or no difference in how he views his experience with the automated equipment. Printers who expect to retire within the life of the contract (mean age: sixty-two) and those who do not (mean age: forty-four) agree at almost the same rate that, compared with the old methods, working with the new equipment is physically easier, intellectually more challenging, and more interesting. In fact, the older men have found their new work physically easier at a slightly higher rate than the younger printers.[5]

Percentage Who Have Found Their New Work	*Younger Men* ($N = 139$)	*Older Men* ($N = 83$)
Physically easier	56	66
Intellectually more demanding	52	56
More interesting	53	46

In the next section we will see that many of the older printers who have not yet been retrained prefer not to get involved in retraining—they would rather sit out the remaining years until retirement. Among older printers who have been retrained, however, many feel that the transition has not been difficult. As one operator, in the trade almost forty years, said, "I'll tell you, it's not hard once you learn it—I like the typewriter. I've found it a challenge, and I think I've picked it up very well. Besides—the atmosphere is better—it's cleaner and quieter."[6]

This was confirmed by several supervisors (foremen) when we asked them whether older printers had difficulty learning the new techniques. On the whole, foremen said that this was not at all the case:

> As a group, the older men seem to be quite enthusiastic about automation. Maybe because they've been through more changes in their lifetime, they're able to handle this one better. After all, they've seen the changeover from manual to tape [teletypesetter] production, and they were around to see the first introduction of cold type. So they're used to seeing changes here.

Said another supervisor, "Overall, the men seem to like the new equipment—the older workers just as much as the younger ones." One foreman vigorously denied our suggestion that the older printer might not take to the new processes as readily as the younger one:

> No—there's no difference! It's a matter of aptitude and personality. Some of our men in their seventies are very much into cold type. In fact, we have many older men with the energy and freshness of a forty-year-old and some forty-year-olds who are like men of seventy.

Most laymen, and many professionals as well, take as given that older workers will not adapt as readily as younger ones to technological change. For example, two contributors to the *Handbook of Social Gerontology* (Slavick and Wolfbein 1960) note that advances in technology "may make long-held skills obsolete and compel the older worker to seek new employment in competition with younger . . . employees *who are more adaptable to the new techniques*" (emphasis added).

When appropriate methods of training have been developed, however, older workers generally achieve results that compare favorably with those of younger persons (Belbin and Belbin 1968; Mullan and Gorman 1972). In fact, studies show that older workers often have *better* work records than the young—less absenteeism, lower turnover, steadier productivity rates, and even higher-quality work (Butler 1975). Our data provide further evidence that when older workers are given the opportunity for retraining, not only can they learn new methods, but they are also almost as likely as younger ones to enjoy their new work as much or more than the old way.

Still, the stereotype remains of the older worker resisting change, unable to learn new techniques, costly and problematic to retrain. And this stereotype persists among some in the ranks of management at the newspapers. One production manager told us, "The young men are taking to automation quite well. But the older ones resist it. They like the old ways, and it's hard for them to see their world changing."

True, among those who have not yet been retrained, more older printers than not prefer to remain with those old methods until retirement. But among those who have been exposed to the new processes, age makes little difference in how these men compare their new work with the old methods.

In 1978 Congress voted to extend the mandatory retirement age from sixty-five to seventy. There is little agreement on what proportion of older workers will exercise the option of remaining in their jobs past the age of sixty-five. On the one hand, more workers than ever before have taken advantage of early retirement incentives and are opting for a life of leisure well before they reach sixty-five. On the other hand, double-digit inflation—and predictions of more of the same—may keep many more people at their jobs until they are forced to retire at seventy, thus avoiding the problem of trying to live on a fixed retirement income.[7]

Over the next decade, then, the U.S. workforce may become an older one, while at the same time technology continues to alter the nature of the workday in many industries. Will the older worker be able to adjust? Our data indicate that, given the opportunity for retraining, many older workers—from both their own and their supervisors' point of view—can successfully adapt to new techniques at the workplace.

The difference in the response to working with the new equipment rests more than anything else on a printer's job classification, for this largely determines to what phase of the process he is assigned. The machinists, who are now servicing the new equipment, are the most enthusiastic. Eight out of ten find their new work both more challenging and more interesting than the old ways.

As a result of automation, the machinist is no longer the "low man on the totem pole."[8] Many have been retrained as technicians and now have the respect of their fellow workers as they exhibit the knowledge and skill involved in servicing the new, expensive, and complex machinery. It is hardly surprising, then, that almost to a man the machinists have responded positively to their new work.

A far different response is that of the proofreaders—two out of three feel that working with the new equipment is *not* intellectually more challenging, and almost all find the new work *less* interesting.

Proofreaders have traditionally viewed themselves as the elite of the composing room. Today it is different: "The proofreader doesn't have the importance he did before automation. Soon all the writers will be preparing their own copy in such a way that the skill of the proofreader will no longer be needed." One source of satisfaction commonly expressed by the men in the proofroom is the ability to "catch the errors" which have escaped the copy editor. With automation, this challenge is gone. Said one reader: "In this room you had to have a certain education—be able to spell and to know your grammar. I liked the feeling of catching as many errors as possible. It's a lot less interesting now because all I do is read a computer printout with a lot of symbols."

Automation has radically altered the work of the proofreader. Instead of correcting editorials and stories, straight from the newsroom, proofreaders receive computer printouts with symbols which must be scanned and corrected. Many find this work not only boring, but also a strain on the eyes. In fact, one in three proofreaders, compared with one in six of all printers who have worked with the new processes, have found their work physically *more* demanding. Further, as the amount of material requiring the old proofing methods diminishes, at one paper many proofreaders have been assigned to "billing." This routine work is far removed from the old excitement of "being on top of the latest news": "Before I used to deal with actual proof. I'm doing billing work, while I used to make printer's marks and corrections. It's taken most of the pleasure away. I've become less of a printer and more of a glorified clerk." Thus, proofreaders are hardly favorable as they compare the new methods with the old and are most likely to take a dim view of automation.

About one-half of both the operators and the handmen think that their new work with the automated equipment is more interesting than the old processes. However, handmen are far more ready than operators to agree that working with the new equipment is intellectually more challenging. And there is good reason for the difference. Operators, as a rule, have been retrained for the new keyboard processes (the Selectric typewriter or the VDT), while handmen generally work at nonkeyboard operations (paste-up). Most of the men, including the supervisors, agree that the keyboard operations are boring: "After all, typing—that's really women's work [sic]. I don't like typing or the whole set-up where you sit in a room with only three or four people." "You do your ads on the machine, and it's like looking at TV for six and a half hours a day. You're not working with your hands or using your imagination."

Several handmen, in contrast, noted that considerable skill and even artistic ability are demanded in the process of pasting up advertising copy for reproduction of the page. As one told us, "It takes a lot of creative skill—layout is just like one of the creative arts. It's a challenge because you have to concentrate on what you're doing." As a result, handmen are more likely than operators to view their work with the new equipment as intellectually more challenging than the old processes.

The differences between the operators and the handmen, however, do not affect their overall appraisal of work. An index combining the responses of each printer to various indicators of job satisfaction was constructed.[9] Somewhat more than 44 percent ranked high on this index; 28 percent, medium; and 28 percent, low.

Despite the fact that operators are less likely than handmen to have found their new work intellectually more challenging, the overall satisfaction level of the two groups is similar. Apparently, after weighing the lack of intellectual challenge against lifetime job security and guaranteed cost-of-living increases, many operators have come to terms with the Selectric typewriter and the video display terminal.[10]

Percentage Ranking High on Index of Job Satisfaction

Retrained operators	47	(74)
Retrained handmen	40	(114)

In sum, three out of five printers have been retrained and are now working with cold type. While management has tended to select the younger men for retraining, many of the older printers are also operating the new equipment and have made the transition with relative ease. They find their new work more challenging and interesting than the old methods at about the same rate as their younger coworkers. Neither the length of time a printer has been working with cold type nor his educational background predicts his response to working with the new equipment.

Only whether a printer is an operator, a handman, a proofreader, or a machinist makes a difference in how he feels about working with the new equipment. The majority of handmen and operators think that working with cold type is more interesting than working with hot metal, although handmen are more likely than operators to feel that the new work is intellectually more challenging. Only a small minority of printers have found working with the new equipment *less* interesting or challenging.

Most of the printers in the composing room make a distinction between how they feel about working with the new equipment per se and how they feel about their job in general. With the exception of the proofreaders, overall job satisfaction is not colored by whatever reservations printers may have about working with the new technology. Despite some specific complaints about the new technology and the feeling that photocomposition lacks the challenge of the hot-lead process, still most retrained printers in the composing room like their jobs. As one summed it up, "Sure I liked the old way better. But still, I have to admit it's a sweet [easy] job now and I like it!"

The Printer Who Has Not Been Retrained

Side by side with those working with cold type in the spring of 1976 were 40 percent of the printers at the three newspapers who had not as yet been retrained. A little more than one-half say they would like to use the new equipment, two out of five say no; and a few feel that it depends on whether they are still around when their turn comes or whether everyone else is using it. The men who would like to be retrained strongly resemble those whom management has already selected to learn the new techniques. They are younger men, at least high school graduates, and they aspire to supervisory positions. A considerable proportion of the older printers—especially those who plan to retire before the expiration of the contract—are opting to "sit it out." Just two-fifths of those planning to retire within the life of the contract (by 1985), but four-fifths of those who do not, would like to be selected for retraining.

Three out of five of those still working with hot metal anticipate that automation will make their job physically less demanding. One-half expect their jobs to become intellectually more challenging, and the same proportion think that, compared with what they are now doing, working with the new equipment will be more interesting.

On what do these men base their expectations? In the relaxed, informal atmosphere of the composing room, the "word" about automation is spread directly from those with experience to those who have yet to be retrained. While the men may no longer spend time together after work, they do converse and visit with one another in the shop, eat lunch together, and play cards or chess in the chapel office or in a corner of the composing room. Shop talk is so common that the printer sometimes preceded his answer to our questions with "Well, a lot of the guys say . . . [it's more/less interesting, and so on]." At two of the papers printers also see their coworkers operating the new equipment. At the *Daily News* this equipment is on the floor of the composing room, separated by glass partitions from the old equipment. At *The New York Times* some trainees are located on a different floor, but generally men work on the new VDTs side by side with those who are composing advertisements by hand, although out of sight of the proofreaders and linotype operators. At the *New York Post*, no new equipment can be seen in the composing room; the trainees are separated from the rest of the printers on another floor, and in fact during their training period they are moved from the night or lobster to the day shift. So visibility of the new operation is highest at the *Daily News*, somewhere in between at *The New York Times* and lowest at the *New York Post*.[11] If the congruence of the evaluations of those who have and those who have not worked with the new equipment rests in part on the visibility of the new machinery, then we would expect this congruence to be the highest at the *Daily News* and lowest at the *New York Post*, and indeed this is the case. The least congruence between experience and expectation is at the *New York Post*; the highest congruence is at the *Daily News*, with *The New York Times* in the middle.[12]

The desire for retraining rests on the expectation that automation will enhance the work of the printer. Those who anticipate that the cold-type process will make their work intellectually more challenging or more interesting are more likely than those who do not to look forward to using the new equipment.[13] Reluctance to be retrained, however, does not stem from a fear that working with cold type will be physically more demanding. Printers who expect that the new processes will be physically more exerting opt for retraining at a rate of 59 percent.[14]

Think Working with the New Equipment Will Be:	*Percentage Who Want to Use the New Equipment*
Physically easier	54 (89)
Physically more demanding	59 (27)
Intellectually more challenging	62 (78)
Intellectually less challenging	36 (42)
More interesting	71 (80)
Less interesting	29 (51)

Recall that among the men who have already been retrained, age made little or no difference in how they appraised working with cold type, as compared with the old methods. This is not the case, however, among the men yet to be retrained. Along with older printers' reluctance to be retrained goes the expectation that, compared with hot metal, the cold-type method will be less interesting and less intellectually challenging. For example, men over sixty are more than three times as likely as those under forty-five to feel that working with the new technology will be less challenging than working with hot metal. Similarly, the older men are considerably more likely than the younger ones to expect that working with cold type will be less interesting than the old way.

Some rationalizing is probably going on here. On the whole, these older men do not want to be retrained. Many feel, too, that management would rather not invest in retraining them. Accordingly, they have adjusted their expectations about the new processes to mesh with the reality that, more interesting or not, *they* will not be learning how to operate the new equipment.

Just as with the printer who has been retrained, the kind of work a printer does—and consequently the kind of work he expects he will be doing after retraining—is the key to how he views the prospect of working with the new equipment. The expectations of operators, handmen, printers, and machinists who have yet to be retrained closely parallel the experiences of those working with cold type. For example, all but one of the seven machinists who have not been retrained want to learn the new processes, and all seven think that working under automation will be more interesting.

The proofreaders are the least enthusiastic about learning the new techniques. Fewer than two in five (38 percent) look forward to retraining, and of all the men they are least likely to anticipate that their work will be more interesting with automation. In fact, three out of five anticipate that it will be physically no easier, and more than one-third expect that automation will make their job less interesting. The strain on their eyes will be greater, and proofing computer codes is dull compared with correcting newsroom galleys.

In some respects, operators and handmen concur about the prospect of working with the new equipment. Two out of three agree that their work will be physically easier, and about one-half of both groups think that working with the new equipment will be more interesting. However, handmen more than operators look forward to retraining and are almost twice as likely (65 percent compared with 34 percent) to think that working under automation will be intellectually more challenging.

As with their retrained counterparts, the nature of the tasks which operators and handmen expect to perform in an automated setting accounts for the difference in their respective stances toward the new equipment. Just as the retrained operators and handmen differ in their assessment of the intellectual challenge of their new tasks, so do operators and handmen who have yet to be retrained. Handmen may be more enthusiastic than operators about the prospect of working with the new equipment because they anticipate assignment to a phase of production which promises a fair degree of skill and challenge; operators, in contrast, see themselves sitting in front of a typewriter day after day, a tedious prospect.

The differences in the responses of operators and handmen to the prospect of working with cold type are, in part, a reflection of their *present* degree of satisfaction, that is, how interesting and challenging they find their current work. Operators take a more negative view of the intellectual challenge of automation because, compared with the handmen, they find their current work sufficiently challenging. Among operators still working with hot metal, two out of five, but just one in four handmen, feel that their current work challenges their ability. Similarly, two-thirds of the operators, but just one-half of the handmen, find working with hot metal "really interesting." Good reason, then, that operators opt at a higher rate to sit it out until retirement. From their perspective, the linotype looks good when compared with the prospect of spending the workday at a typewriter.

Summary

Two-fifths of the printers have yet to be retrained. Of this group, the younger men with more years of schooling, and with an eye to promotion, look forward to retraining while their older coworkers opt to watch the automation process from the sidelines. Those who anticipate that working with cold type will be intellectually more challenging and interesting want to learn how to operate the new equipment. But those who have defined the new work as less intellectually challenging and interesting than what they are now doing prefer not to be retrained.

Further, the new technology is viewed differently by men yet to be retrained, depending on their occupational classification. Machinists are most, and proofreaders least, enthusiastic about working with the automated equipment. The differences in response to the prospect of working with cold type depend both on the printer's expectation of the type of work for which he will most likely be retrained and on his satisfaction with doing a day's or night's work the old way.

The varying expectations about what working in the cold-type composing room will be like are anchored in reality, for they mirror the actual response of those who have already been retrained. The yet-to-be-retrained proofreader sees his fellow readers bent over computer copy or doing billing work, while the machinist observes his retrained counterpart's new, enhanced status of technician. Similarly, operators and handmen who are still working with hot metal observe and hear about the new tasks to which retrained operators and handmen have been assigned.

On balance, the majority of men who have been retrained feel that their new tasks are as challenging and interesting, if not more so, than the old methods. And among the group yet to be retrained, most agree that working with the new technology will be at least as satisfactory as working with hot metal. Job satisfaction is still high, and dissatisfaction is in evidence only among one small group—the proofreaders. The men who have been retrained have adjusted to their new work, and most of those who have yet to be retrained anticipate that they, too, will adjust. In other words, these printers have largely come to terms with the technological changes in the composing room.

Still the question remains: How have these printers managed to come to terms with the sweeping changes which have altered, or are expected to alter, the nature of their workday?

In the next chapter we continue our account of the printers' response to automation by classifying printers into two types: those with a primarily instrumental and those with a primarily traditional orientation toward work. We then examine the ways in which these differing orientations have enabled the printer to accept the changes in the composing room.

Notes

1. For many years Local 6 had resisted management attempts to assume control over the introduction of automation into the composing room. But it became clear, when the previous contract expired in 1973, that

the very survival of the newspaper industry in New York City depended on the ability of the publishers to automate. Hard bargaining enabled the union at least to ensure—through the lifetime job security clause—that no printer would lose his job as a result of automation.

2. Among men on the night shift, 54 percent have been retrained, compared with 63 percent on the day and 68 pecent on the lobster shift.

3. It could be that selection for retraining itself stimulates the aspiration for promotion. However, a number of men who had not yet been selected indicated that the reason they wanted to be retrained was that "it [retraining] is the only way to get ahead here."

4. Throughout the analysis we have assumed that the response "intellectually *more* challenging" is a positive one, similar to the response "more interesting." Although printers are technically blue-collar workers, they have always considered themselves members of a literate and "intellectual" occupation. More than one printer, in the course of conversation, referred to printing as his "profession." It is quite likely, therefore, that the printer would consider intellectual challenge to be a job asset, rather than a liability. Further, there is a high correlation (gamma = .64) between having found the new techniques more intellectually challenging and more interesting, and the latter is undoubtedly regarded as an asset of the job (*Work in America* 1973).

5. In the remainder of this and the next chapter, unless otherwise specified, those who expect to retire before the contract expires are referred to as the older men, and those who do not as the younger men.

6. On an earlier field visit we had observed this man in the training room, painstakingly searching for the right key, barely managing to type ten or fifteen words a minute. Apparently, once the transition has been made, the problems of the learning process are forgotten.

7. While cost-of-living increases are built into the social security system, this is not so for the newpaper or union pension payments.

8. See note 15, chapter 1.

9. Items included printers' ratings of pay, prestige, job security, interest, challenge of ability, and sense of job well done.

10. Only two of the sixteen proofreaders who have been retrained rank high on this satisfaction index—evidence that for this group the specific changes in their day-to-day work have indeed affected their overall feeling about their job.

11. Visibility may be lowest at the *New York Post* for another reason. Only at this paper have fewer men been retrained than not. At *The New York Times* those with experience outnumber those without by more than 2 to 1 and at the *Daily News* by 1.6 to 1.

12. Since the spring of 1976, when our interviews with the printers were

conducted, all three composing rooms have undergone radical change as the newspapers moved toward completely cold-type production.

13. This is further evidence that most printers feel that greater intellectual challenge is a job asset (see note 4).

14. The causal direction of the differences is difficult to pinpoint. Men who do not want to be retrained may be adjusting their expectations of how the new processes will compare with the old to mesh with their desire to sit it out.

Appendix 2A

Table 2A-1
Who Would Like to Be Retrained? (Percentage Saying Yes)

Age			*Expect to Retire during Contract*		
Under 45	78	(32)	No	81	(52)
45-54	65	(26)	Yes	40	(90)
55-61	50	(22)			
62 +	42	(76)			
Schooling			*Like to Be a Foreman*		
Some college	65	(23)	Yes	75	(36)
High school graduate	61	(69)	No	43	(90)
Some high school or less	47	(57)			

Table 2A-2
Response to Automation by Job Classification

	Operators ($N=69$)	*Handmen* ($N=47$)	*Proofreaders* ($N=17$)	*Machinists*[a] ($N=7$)
Percentage who want to work with the new equipment:	42	66	38	(6)
Percentage who think the new processes will be:				
Physically easier	62	67	40	(5)
Intellectually more challenging	34	65	50	(5)
More interesting	46	51	41	(7)

[a]Too few cases to percentage.

Table 2A-3
Congruence between Experience with and Expectation about Automation by Newspaper

Working with the New Equipment Is (Will Be):	*Daily News*			*The New York Times*			*New York Post*		
	Experience ($N=116$)	*Expectation* ($N=73$)	*Difference*	*Experience* ($N=102$)	*Expectation* ($N=50$)	*Difference*	*Experience* ($N=27$)	*Expectation* ($N=32$)	*Difference*
Physically less demanding	54	60	+6	64	52	−12	52	73	+21
Intellectually more challenging	62	59	−3	42	46	+4	52	38	−14
More interesting	53	53	0	42	46	+4	44	56	+12
Sum of the differences (ignoring plus or minus signs)		9			20			47	

3 Instrumentalists and Traditionalists

From the time of Gutenberg, printers have considered themselves the elite of blue-collar workers, master craftsmen in a trade traditionally passed from grandfather to father to son. Literate when many were not, bound into an occupational community, proud of the prestige of their craft, these men could be expected to resent, if not resist, automation; but they did not. They have come to terms with the technological changes that have been taking place in the composing room.

As we spoke with the printers, it became apparent that underlying their comments were two different orientations toward work, one traditional and the other instrumental (Lockwood 1966; Goldthorpe et al. 1968; Beynon and Blackburn 1972). Specifically, printers were divided between those who took great pleasure in the skills of their craft and those who viewed their work primarily as a means to an economic end.[1] These diverse approaches toward the meaning of work help us to understand the dignity and relative equanimity with which most have accepted the new technology.

Orientations to Work

Several of the questions that we asked printers enable us to identify their orientation as predominantly traditional or instrumental. For example, we asked them what they liked and disliked most about their work. Typical traditional responses focused on the interest and challenge or craftsmanship of the work:

> Every day is a new challenge, with the new ads to make up. I love the excitement of putting the paper to press—that's why I prefer working on the night shift. The work is interesting—not the same old baloney each day.
>
> I love printing; there's nothing I don't enjoy about my work. I'm completely absorbed in it. It's like a musician with a piano—I love to play the keyboard. I can't express it; I just love it.
>
> I've always liked printing. I love getting the ink on my hands. In fact, ink gets into your veins just like an old sailor gets salt in his veins. I have a lot of printer's blood in me. [His brother and son are both in printing.]

Some traditional responses specifically cited automation as a positive factor, enhancing the work:

> Now with the computer, going from hot metal to cold type, every day is even more of a challenge. It's a whole new ballgame and that interests me.

> There are a lot of changes going on in the industry; you know, automation. I feel lucky to be involved in printing as it's going through all these changes. And don't let them tell you that printing is not a craft anymore. It still is—even with automation.

Other traditional responses reflect a negative appraisal of automation:

> What do I dislike? I don't like working in a factory, and that's what it's becoming. The machine does the work, and all you have to do is push a button. It used to be creative, but not anymore.

> Automation is so simple, it takes away a person's incentive. Compared with hot metal the new system is faster, but it's just not the same. Now all you have to do is feed in typed copy from the customer, and that's all there is to it. Maybe someone just starting in will like it, but it's not for me.

Still other responses suggesting a traditional orientation focus on the social context of work, that is, the degree of camaraderie with fellow workers:

> I enjoy the people I work with—they're good to get along with. It's really nice to have fellow workers whom you like and who like you.

> What do I like? The men! We have a lot in common, and we kid around a lot. We all get along well with each other. No bad feelings around here.

In contrast, instrumental responses emphasize economic or favorable situational factors:

> The money is really good—it's a good, steady, secure job.

> The salary—that's the only thing I'm here for.

> The hours are good; I get to work in a couple of minutes and, let's face it—where else can you make this kind of money? I've got a house and two cars.

In the same way, the responses of the printers to other questions were categorized as suggestive of either a traditional or an instrumental orientation. (See appendix 3A for these questions and responses.) By summing their responses to various questions about work, each printer could be given a traditional and an instrumental score depending on the number of answers of each type. Then a typology was constructed—traditional, mixed, or instrumental. A total of 30 percent of the printers were categorized as "mixed"; that is, they gave equal numbers of instrumental and traditional responses. Almost two out of five printers were high on instrumentalism and low on traditionalism, and one in three was the opposite—high on traditionalism and low on instrumentalism. It is these two contrasting groups on which we focus as we examine the printers' response to automation.[2]

The Instrumentalist Printer

John Callahan is a forty-four-year-old operator who has been in printing for twenty-four years, virtually since he finished high school. He has been at *The New York Times* for five years; before that he worked in book and job (commercial) shops. No one in his family was ever in printing, nor are any of his close friends in the trade. He works on the lobster shift and prefers it because "You make more money on this shift; the work is the same on any shift—no difference. Work is work in any language; you do your job; you come and go; and tomorrow's another day." John would not choose printing again if he were starting all over:

> The job market for printers just isn't what it used to be. Much more opportunity in another field. I don't think there's any future for anyone in printing.
>
> What do I like about my work? The salary—the money's very good; I make a good living; that's the only thing I'm here for. I find my work interesting, but it's the salary—that's why I'm here.
>
> There's nothing I dislike about my work except that my slide days [days off] are Tuesday and Wednesday, so I don't have weekends with my family.

John has not taken a sabbatical yet (the contract provides for up to six months of "paid productivity leave" over the eleven-year life of the agreement), but "If I did take a leave, I wouldn't miss my work or the people at the paper. I'd keep busy working around the house." As far as the union is concerned, John does not attend the monthly meetings; but like almost all printers, he is a regular voter in local and international elections; nor has he ever attended any of the clubs or social activities sponsored by the union. However, he thinks the ITU is "the best." "They do very well by us financially. And this last contract with lifetime job security and cost-of-living increases was very good. No, there was nothing I disliked about the contract." He looks forward to retiring at sixty-two because

> It's the earliest time that I can retire and get social security . . . unless they come up with a package that will make it attractive to get out sooner. Why be obligated to a job when you don't have to—it's better to be able to call your time your own.

John has not yet worked with the new equipment, but he would like to:

> The whole country's going automated, and I've got a lot of years yet to go. At least if I learn it [automation], if, God forbid, anything happens here, I can go elsewhere; it's a must!
>
> How do I think I'll like it when the composing room is completely automated? No difference at all. Like I said, I'm just involved in making a living; automated or not—it doesn't matter.

This instrumentally oriented printer is not dissatisfied with his work. He finds it interesting; in fact, he agrees strongly that when he leaves for home, it is with a feeling of a job well done. But he is not interested in any intrinsic satisfactions which he may derive from work. For him, it is a job, and a good one that offers security, relatively high wages, and the promise of a comfortable retirement.

The Traditionalist Printer

Tom D'Amato is a forty-eight-year-old handman, following in the footsteps of his father who was also a printer. He has been working at the *Daily News* for thirty years and loves the atmosphere. "I love the hustle and bustle. That's why I love working the night shift—it's busier on the night side; you have to get the edition out. I like going at top speed to beat the deadline; it's more exciting and challenging." If Tom were starting all over, he would choose printing again.

> Why? I love this life. I love what I'm doing. Every night is a brand new page of history. Never went to high school, but I sure learn a lot here. I really enjoy the work and the people I work with. Like I said, every night is a new challenge to get the paper out and see that the ads get in. I love setting ads and newsmatter—it's an art. It's constructive work, and you feel that you have accomplished something. Also, the relationships are good; there's no bad feelings—no one gets mad at each other. I'll tell you—it's like the daily double here; you got a job *and* you like it!

Tom looks forward to retraining and thinks he will like his work even more once the composing room is completely automated:

> It'll be challenging and more interesting. You'll have to use your head. Some of the fellows don't agree, but I would like to learn something new. When you stop learning, that's a bad sign. The new equipment is very sophisticated, and I think I'll pick it up very well. I'll tell you—it will be a real challenge for a middle-aged printer like me to learn a new craft.

During his eight-week paid productivity leave, Tom kept in touch with things at the paper and acknowledged that he was glad to get back to work. "There wasn't much to do around the house; the old lady works, so we couldn't go places. And I missed the guys here."

The union is an important part of Tom's life. At one time he was a chapel officer, and he still attends many of the monthly meetings as well as some union social activities. In fact, his free time is spent more with other printers (mostly from the same newspaper) than with people in other lines of work. His attitude toward the union is positive:

> The ITU—it's not only the best, it's the greatest. Every union and every industry has its problems. We're all being automated, and under the circumstances the union is doing very well. You know, it's the most democratic union in the country—at least here in New York City it is. Big Six is right up there at the top—it's done well for all of us.

Tom's attachment to his work is summed up in his attitude toward retirement. He expects to keep working as long as he possibly can: "I want to stay here till I die— this is my life!"

Two printers—both interested in their work, both expressing a desire to be retrained, both stout defenders of the union—are very different in their orientation toward work. One, the instrumentalist, leaves his job behind him when the shift is over; for the other, the traditionalist, the job is his life.

Sources of Instrumentalism and Traditionalism

We expected that age would be the strongest predictor of a printer's orientation. After all, the older printer, with long years of experience in the trade (and often with a father or uncle who was also a printer) would probably have developed a strong sense of craft, a close feeling of identification with fellow printers and with the union, and a deeper pride in the skills of his occupation. In contrast, the younger man, with more schooling, with children to feed, clothe, and educate, with mortgage and automobile loan payments to be made each month, as well as with fewer years in which to have developed a sense of craft and occupational identification, might view his job primarily as a means of obtaining those material rewards that will permit him to enjoy a "good life" apart from work. Among printers, however, age does not distinguish traditionalists from instrumentalists—older printers are as likely as younger ones to have an instrumental orientation.

Beynon and Blackburn (1972, p. 150) found that in contrast to those who were single, the commitment to work among family men in the factory they studied was "distinctly instrumental." This was not so among printers.

The single men were as likely as those who were married, and those with younger children only a shade more likely than those without, to be instrumentally oriented.

The only background factor at least somewhat predictive of a printer's orientation is the number of years of schooling he has had. Even here, however, differences are slight, although printers with a high school diploma are less likely than those without one to be traditionalists.

The work situation itself may contribute to the formation of a particular orientation (Beynon and Blackburn 1972, p. 158), and for the printer this means the newspaper itself and his job classification. Between the *Daily News* and *The New York Times* differences are minimal—about one-third of the men at both the *Daily News* and *The New York Times* are instrumentalists; but at the *New York Post*, instrumentalists comprise more than one-half of the composing room force.

The relative insecurity of the printers at the *New York Post* probably explains the predominance of instrumentalists there. The publisher was far from young, business was not good, and many men were concerned about their "lifetime job guarantee" in the event of the closing of the paper. With economic fears paramount, pride of craft or loss of long-held skills is less salient. And so their answers focused on the extrinsic aspects of the job, on the economic (rather than social or political) successes or failures of the union, on the material benefits or drawbacks of the new contract, all of which contributed to higher instrumental and lower traditional scores at the *New York Post*.

Another aspect of the work situation—the printer's job classification—differentiates instrumentalists from traditionalists. Among operators and handmen, little difference in orientation exists; two in five are instrumentalists, and about one-third are traditionalists. The difference in the nature of the tasks performed by proofreaders and machinists, however, may account for the fact that one-half of the proofreaders, but only one-fourth of the machinists, are traditionalists. Proofreaders—at least in the hot-metal composing room—have long taken pride in the degree of skill and accuracy required to correct copy. This group has always considered itself the elite of the composing room.[3]

Machinists, on the other hand, are responsible for servicing the equipment in the shop, a function which hardly makes them skilled craftsmen in the eyes of their fellow printers.[4] Nor do machinists themselves feel as strongly as the other men in the shop about the prestige of printing.[5] The difference in the orientation of proofreaders and machinists highlights how divergent the work values of men who work in the same setting may be. All the same, both have come to terms with technological change.

Instrumental and Traditional Responses to Automation

The Retrained Printer

Traditionalists and instrumentalists have been retrained at approximately the same rate: 61 percent of the former and 57 percent of the latter are now working with cold type. While we might have expected the traditionalist, with his strong sense of craft, to less satisfied working with equipment which has rendered his old skills obsolete, this is not necessarily the case. Among the men who have been retrained, orientation does not predict how interesting or challenging printers have found the new processes compared with the old. Almost one-half of both the instrumentalists and the traditionalists have found working with cold type more intellectually challenging and more interesting.

However, the consensus is more apparent than real. Distinct differences between the two groups appear if we look separately at the younger men who expect to remain in the composing room after the contract expires and the older ones who will be retiring in the next few years. For example, among the younger men, traditionalists are about equally likely, if not more likely than instrumentalists, to find working with the new processes more interesting. In contrast, the older traditionalists are considerably less likely than older instrumentalists to think that their new work is more interesting.

	Percentage Who Have Found the New Processes More Interesting			
Orientation	*Younger Men*		*Older Men*	
Instrumentalists	46	(55)	63	(27)
Traditionalists	53	(47)	38	(24)
Difference	+7		−25	

The older traditionalist looks back with nostalgia at the old, familiar methods from which he derived pride and satisfaction and has found the new processes wanting in challenge and interest: "I loved what I was doing before. I'm a man who likes to work with my hands, and now my hands are tied."

The younger traditionalist, looking ahead to many years of working in a cold-type composing room, has transferred his tendency to seek intrinsic satisfaction from work from the old to the new cold-type processes.

> The new equipment is very sophisticated, and I found I moved right on to it. I enjoy the challenge of learning something new. You have to use your head.
>
> I've always been fascinated by electronics, and now I have the chance to learn all about it. The old way was fine, but this is really more exciting, more interesting.

Most satisfied with the new methods are the older, soon-to-retire instrumentalists, a large majority of whom say that working with the new equipment is more interesting than working with hot metal. Their comments suggest, however, that they are largely referring to the more satisfactory *conditions of work* under the new processes. For example, they praised the orderliness of the new procedures: "You can only mess up one thing instead of several things." Some liked the more white-collar atmosphere: "I don't have to worry about getting my hands dirty; I can dress in a shirt and tie now." A handman noted: "I like the fact that I can sit down. You get tired of standing while you make up a page with hot metal." Still another said, "It's pleasant—there's not too much pressure. It's quieter, and also cooler, because where I am it's air-conditioned."

Thus the older instrumentalist finds the new processes more interesting not because of the intrinsic nature of the tasks he is performing, but because he finds the conditions under which he is now working more pleasing.

About one-half of the young instrumentalists find their current work more interesting than the old methods. We have no way of knowing how interesting they found working with hot metal, but their comments indicate that how interesting their work is just does not matter. One young operator put it bluntly: "I have to like it because there's no other place to go." A young handman added, "Automated or not—it doesn't make a difference. I'm just involved in making a living."

In general, instrumentalists are more indifferent than traditionalists to the challenge of their new tasks. Almost one in four instrumentalists, for example, compared with one in ten traditionalists, says that as far as interest is concerned, there is no difference between the old and the new methods. Similarly, instrumentalists are more likely than traditionalists to report no differences in the intellectual challenge of cold type compared with hot metal.

In chapter 2 we saw that, with the exception of a small minority of printers, the men who have been retrained have made the transition from hot metal to cold type with little anger, bitterness, or resentment. Some insight into why this has been the case is provided by understanding that these men approach their work from two different perspectives and that each of these perspectives, in its own way, has facilitated the printer's ability to come to terms with the new technology.

Before discussing how these two orientations to work have helped ease the transition to cold type, we turn to the printer who has not been retrained to see whether his orientation to work has influenced his desire to learn the new techniques and his appraisal of what it will be like to work under automation.

The Printer Who Has Not Been Retrained

Overall, the instrumentalist is more likely than the traditionalist to look forward to retraining. Almost three out of five instrumentalists, but fewer than one-half of the traditionalists, would like to work with the new equipment. Two reasons account for the lack of enthusiasm of the traditionalists. First, they are satisfied with their work in hot metal. Far more than the instrumentalists they find their work "really interesting," they leave the composing room with the feeling of a job well done, and they consider their trade prestigious.

Percentage Who:	*Traditionalists* ($N=51$)	*Instrumentalists* ($N=66$)
Find the job really interesting	71	50
Strongly agree that their work gives them a feeling of a job well done	45	26
Rate the prestige of printing "excellent"	41	14

Second, the traditionalists do not think that working under automation will compare favorably with their present work. More than the instrumentalists, they fear that the new work will be physically as or more demanding and less interesting.

Percentage Who Believe Working under Automation Will Be:	*Traditionalists* ($N=47$)	*Instrumentalists* ($N=63$)
Physically as or more demanding	49	34
Less interesting	45	28

We already know that younger men are far more likely than older ones to look forward to retraining. A separate look at the responses of instrumentalists and traditionalists who expect to retire within the life of the contract and those who do not reveals that younger men are more than twice as likely as the older printers to look forward to learning the new processes. Among both these groups, however, an instrumental orientation still yields a somewhat higher proportion opting for retraining.

Age is clearly the key to the desire for retraining, but the printer's orientation tells us *why* the older man resists and the younger man looks forward to the opportunity to learn the new processes. The difference between the older and younger traditionalists and instrumentalists emerge graphically as they respond to the question Why do (don't) you want to work with the new equipment?

The Older Printer. Of the older traditionalists 70 percent prefer not to get involved with automation. Their reasons are that they are used to and like the old ways better: "Having worked forty-six years doing what I'm doing, I'm set in my ways; my first love is linotype operating." Or, "I'm from the old school. I'm not interested in new ways—I love the old printing. Hot metal has a lot more romance to it." One older traditionalist suggested that automation destroys both the craftsmanship of the printer and the quality of the product: "I think the new equipment is taking the skill out of work. A machine tells you what to do. Besides, you get better quality the old way." The exception is the older traditionalist who wants to be a part of the new technology: "I don't think anybody should stop progress; I'd love to see what it's all about—get my finger in the pie."

On the other hand, the older instrumentalists saw no reason to take on the responsibility of learning something new at their stage of life. Why should management invest in retraining them when retirement is just around the corner? In one form or another they said, "I'm just waiting for my time; as soon as they give me a good incentive, out I go." And ". . . I expect to retire in six months; it doesn't pay for them to train me."

Their reasons hardly suggest that these skilled craftsmen, who have been in the trade an average thirty-five years, are alienated or even concerned about the coming obsolescence of their skills. Most just do not want to be bothered learning new skills at their age. Their jobs are secure; the wages are good. Why get involved?[6]

The Younger Printer. More than three in four of the younger printers are enthusiastic about the prospect of the retraining, but for different reasons. The traditionalist feels that his work will become more interesting and challenging.[7] "I've always loved electronics. It'll be exciting. Automated processes are more demanding intellectually."

The younger instrumentalist, in contrast, is eager to learn the new processes. His reasons, however, have little to do with the intrinsic aspects of the work but focus rather on the fact that he wants a job—today and in the future. "I have to learn this process, or *I'll* be obsolete. Besides, it means more job security." Or, put more graphically, "I want to be where it's at for tomorrow—I don't want to push a broom."

The instrumental orientation is perhaps best captured in the words of one younger printer: "I'll like it as long as they pay me!" The outlook of these younger instrumentalists is clear and unequivocal: The route to economic security and promotion lies in retraining—and, if at the same time it is interesting, so much the better.

In sum, when we look at older and younger printers with different orientations, we find that age determines the direction of a given orientation. The older traditionalist looks back; he likes the old way, prefers to cling to the old linotype or proof symbols. The younger traditionalist looks ahead; he has switched his concern for the intrinsic factors to the new methods and speaks of the challenge of electronics and the thrill of learning new techniques. For the older instrumentalist it is just too late—"no point in getting involved at this stage." And for the younger instrumentalist there is no choice—"It may or may not be more interesting, but that's progresss."

By the time the composing room is fully automated, the older printer, whether traditionalist or instrumentalist, will be retired or about to retire. Remaining will be the younger men. According to management, the printers will have been retrained in both keyboard and nonkeyboard tasks, and some of the differences we found in response to automation between machinists and proofreaders, or between operators and handmen, will be history. (See chapter 2.) More critical is the orientation toward work which will, in all probability, predict the morale in the composing room of the future. To this point, the prognosis is positive. The younger traditionalist, with only a minority of dissenters, expects to find, or has already found, intrinsic satisfaction in the challenge of the new electronic processes. The younger instrumentalist has adjusted, or expects to adjust, positively to his new work regardless of the degree of intrinsic satisfaction it may provide. The extrinsic satisfactions are all there—good pay, decent benefits, job security, and a fairly comfortable retirement income to look forward to.

The Composing Room of the Future

As a check on how printers see their future, we asked all of them how they expected to like working in a fully automated composing room Almost one-half of the men said they would like it more, one in four would like it less, and the rest felt it would make no difference or would depend on what the new setup would be like. The younger printer is considerably more likely than the older one to view the prospect of a fully automated composing room with equanimity, and this is true whether or not he has had experience with the new equipment. Retrained or not, about three out of five of the younger men think they will like their jobs more. Among the older printers, however, experience makes a significant difference in how they feel about the workplace of the future. The retrained, older printers are considerably more likely than those who have not worked with the new equipment to think they will like their work more once automation of the composing room is complete. Thus, not only have these older workers been successfully retrained (see chapter 2), but also once they have settled down to working with the new technology, they like it.

Percentage Who Expect to Like the Composing Room More When Fully Automated

	Retrained	*Not Retrained*	*Difference*
Older men	43 (75)	27 (86)	+16
Younger men	58 (133)	57 (49)	+1

Orientation and the Future

Is the prospect of working in a fully automated composing room viewed differently depending on whether a printer's orientation toward work is instrumental or traditional? At first glance the answer is no. Traditionalists are slightly more likely to say that they will like the composing room more once it has been fully automated, but at the same time they are more likely than instrumentalists to say that they will like it less. However, where traditionalists and instrumentalists diverge is in the middle category, the "it makes no difference" response. Printers with an instrumental orientation toward work are considerably more likely than their traditional fellow workers to feel that "it just won't make a difference." Hot metal or cold type, "work is work; do your job and get it out!"

The tendency for it not to make a difference is greater for instrumentalists than for traditionalists whether they are older or younger men and whether or not they have had experience with the new equipment. In every instance, but particularly among printers who have been retrained, instrumentalists are more likely than traditionalists to shrug and say "no difference."

Percentage Saying "No Difference"

	Instrumentalists	*Traditionalists*
Older men	37 (54)	25 (48)
Younger men	39 (72)	21 (63)
Retrained	44 (73)	21 (67)
Not retrained	30 (53)	25 (44)

Further, the instrumentalist's attitude toward the composing room of the future is hardly affected even by how interesting he has found the cold-type as compared with the hot-metal process. For example, whether the retrained instrumentalist thinks that working with the new equipment is more or less interesting makes very little difference in how he views the prospect of working in a fully automated composing room. This is not so for the retrained traditionalist. Almost none who has found his work less interesting, but almost every traditionalist who has found working with the new equipment more interesting than hot metal, looks forward to his job in the fully automated composing room.

The same can be said of the printer who is still working with hot metal. The degree of intrinsic satisfaction that the traditionalist expects to find in working with the new equipment makes a sizable difference in how much he thinks he will enjoy his work in the fully automated composing room. Whether the instrumentalist, in contrast, thinks that working with the new equipment will be more or less interesting is irrelevant.

	Percentage of Retrained Printers Who Expect to Like the Fully Automated Composing Room More	
Has Found Automation	*Instrumentalists*	*Traditionalists*
More interesting	62 (34)	85 (34)
Less interesting	44 (27)	19 (31)
Difference:	+18	+66

Further, even if the instrumentalist thinks that working with cold type will be *less* interesting, his overall attitude toward working under automation is, at worst, indifferent. More than four out of ten instrumentalists who anticipate that working with the new equipment will be less interesting and less intellectually challenging said that, as far as liking their jobs once the automation process has been completed, it will make no difference. Traditionalists are far less likely to feel this way. For them, the degree of interest which they anticipate in working with the new equipment does make a difference: If they think their work will be less interesting, they expect that they will like their job *less* once the composing room is completely automated. If, on the other hand, they think working with the automated equipment will be more interesting, then they are quite optimistic about the prospect of working in the fully automated setting.

The difference between traditionalists and instrumentalists, in other words, is not whether they look forward to the fully automated composing room. Age and experience are the key to whether printers take an optimistic or a pessimistic stance toward the composing room of the future.[8]

Regardless of orientation, younger printers and those who have been retrained look forward to the fully automated composing room considerably more than do the older printers and those who have not been retrained.

What orientation tells us is *why* these men—old or young, retrained or not—are, on the whole, coming to terms with the sweeping technological changes taking place at the workplace. In a word, everyone recognizes the inevitability of automation. The instrumentalist, however, shrugs it off. The traditionalist, perhaps rationalizing, foresees that his new work will provide intrinsic satisfactions comparable to those he derived from the old skilled craft of printing.

Summary

Here, then, are men who see their world changing. No longer will they be operating the linotype machine, skillfully setting type by hand, or carefully correcting fresh copy as it arrives from the newsroom. And most are not bitter, resentful, or angry. The older printers know that they will be retiring soon. Some traditionalists among them may resist retraining because they feel the new methods lack challenge, but the majority are more instrumental in their attitude. They simply feel that at their age, it does not pay, from either their own or management's point of view, to learn the new processes.

As for the younger printers, most have been retrained, and almost all who have not are looking forward to working under automation. Only a few of the traditionalists among them have found their new work less interesting and are nostalgic about the old linotype machine, the grease, and the smell of printers' ink. The majority have found their new work equally or more interesting and challenging than the old hot-metal process, and they look forward to the time when the composing room will be fully automated. The younger instrumentalists also see that the future is automation, and they want to be part of it. They do not think about liking or not liking their new work. Cushioned by job security and a good salary, they have, in a manner, shrugged off the loss of craft and turned to their new tasks—some with enthusiasm, others with indifference. Thus the printers have taken technological change in stride, continuing to do their day's work with dignity and a sense of worth.

In 1976, of course, there still was a day's work for every printer who chose to remain in the composing room. At question, however, is whether, once the papers have gone completed cold type, there will still be work for every printer. Management and some of the printers themselves are seriously concerned about this eventuality. As automation progresses, there will be less and less to do, and fewer and fewer men will be kept busy. The printers have adjusted well to the loss of craft, but even the instrumentalists among them may have difficulty handling the loss of work. These craftsmen will not be satisified to collect a weekly paycheck for sitting around and playing cards. But few look so far ahead at this time.

Notes

1. See the last section of chapter 1 for a discussion of these two orientations.

2. We have identified them as instrumentalists and traditionalists, but the reader should bear in mind that they represent not pure types, but rather approximations of two different orientations toward work.

3. This explanation receives empirical support when we note that the instrumentalists among the proofreaders are largely those who have already switched over to cold type and are deciphering computer symbols rather than reading proof.

4. In fact, they are hardly regarded as printers by the other men in the composing room (see chapter 2).

5. Machinists were only one-half as likely as the operators, handmen, and proofreaders to rate the prestige of printing as excellent.

6. It has been suggested that the reluctance of some older workers to be retrained reflects a lack of confidence in the ability to learn new skills at their age (Belbin and Belbin 1968, p. 344). The comment of one—"I'm too old for all that retraining; I'm on my way to the farm"—reflects this view.

7. Only five of the eighteen younger traditionalists agree with their older counterparts and complain: "There doesn't seem to be any challenge in the computer; and the more the craft part of it goes, the less interesting it will be."

8. Regressing this variable on age, experience, schooling, job classification, and so on showed that age and experience account for almost all the explained variance in printers' attitudes toward the composing room of the future.

Appendix 3A

Table 3A-1
Responses Suggestive of a Traditional or Instrumental Orientation

Question	*Traditional Orientation*	*Instrumental Orientation*
What are some of the main things you like about your work?	Interesting, creative Just love printing, enjoy everything Challenge (learn new skills) Fellow workers; camaraderie	Money, benefits Security Hours, convenience (travel) Can wear shirt and tie; clean, more like white-collar job
What are some of the main things you dislike about your work?	Boring, repetitive Not a craft anymore	Money, benefits Hours, inconvenience
Why (if you were fifteen and starting all over) would you choose printing again?	Creative, interesting Love printing craft	Money, benefits Difficult to get into anything else
Why (if you were fifteen and starting all over) would you choose another occupation?	Craft is changing Less interesting or creative	No opportunity in printing Better money in another occupation
Why would you like to work with the new equipment?	It will be interesting, challenging	No choice, it's the future Only way to get ahead Doesn't make a difference; a job is a job
Why would you rather *not* work with the new equipment?	Like what I'm doing now Not a craft; woman's work Boring, routine	Too old; wouldn't pay; Doesn't make a difference; a job is a job
What are the main things you like about the new contract?	Automation, new work Training possibilities	Security (lifetime guarantee) Cost-of-living increases
What are the main things you dislike about the new contract?	Management control over my work	Money, fringe benefits

Table 3A-2
Printers' Orientation by Age, Marital Status, and Family Composition
(percent)

	Age			*Married*		*Child under 18*	
Orientation	*Under 45*	*45-59*	*60+*	*Yes*	*No*[a]	*Yes*	*No*
Instrumental	38	39	39	38	39	41	34
Mixed	28	31	32	30	29	26	34
Traditional	34	30	29	32	32	33	32
(*N*)	(131)	(119)	(143)	(340)	(68)	(182)	(188)

[a]Includes men who are widowed, divorced, or separated.

Table 3A-3
Printers' Orientation by Newspaper and Occupational Classification
(percent)

Orientation	*Daily News*	*New York Post*	*The New York Times*	*Oper-ator*	*Hand-man*	*Reader*	*Mach-inist*
Instrumental	35	52	38	42	39	29	53
Mixed	30	26	32	26	33	24	12
Traditional	35	23	31	33	28	47	35
(*N*)	(192)	(62)	(154)	(147)	(166)	(34)	(17)

Table 3A-4
Percentage of Printers Who Want to Use the New Equipment

	Traditionalists	*Instrumentalists*
Younger men	72 (18)	83 (23)
Older men	30 (27)	39 (33)

Table 3A-5
Attitude of Instrumentalists and Traditionalists toward the Fully Automated Composing Room
(percent)

Percentage Who Will Like the Fully Automated Composing Room	*Instrumentalists* (*N*= 138)	*Traditionalists* (*N*= 123)
Less	22	31
No difference	38	23
More	41	46

Table 3A-6
Percentage of Printers (Not Retrained) Who Expect to Like the Fully Automated Composing Room More

Expects to Find Automation:	*Instrumentalists*	*Traditionalists*
More interesting	41 (29)	57 (22)
Less interesting	12 (16)	14 (28)
Difference:	+29	+43

4 One Printer Retires, Another Does Not

Introduction

The worker who faces mandatory retirement knows that when he reaches a certain age, out he goes. But for members of Local 6 employed at the *Daily News, New York Post,* or *The New York Times,* retirement is voluntary; they are not forced out. These printers have a second advantage. Their union contract guarantees them job security at whichever newspaper they are employed. Except for the close of a paper or a printer's discharge for incompetency or neglect of duty, he may continue where he is. Since no one must seek another job and no one is forced to retire, these printers present an excellent opportunity for finding out about basic attitudes and values concerning work, retirement, and leisure.[1] Under these circumstances, then, of economic security and voluntary retirement, we examine the impact of a dollar incentive to retire.

The 1974 collective-bargaining agreement between the Publishers' Association of New York City and Local 6 stipulated that any printer who retired during the first six months of the contract would receive a $2,500 cash bonus.[2] "A spur to quick retirement" was the way the bonus was hailed,[3] and the research question is, What have the results been? Who decided to take the bonus and who wanted no part of it?

The main reason for comparing the printer who took the bonus with the one who did not is to understand what work and retirement mean to men when they have a choice. Our strategy is to examine some dimensions of their lives—their backgrounds, their work experience, their attitudes toward work versus retirement, and the special features surrounding employment in the composing room of New York City's three largest papers.

The Retirement Incentive

The men who chose to retire knew that they could have continued to work in a place which they liked, which they knew well, where they had friends, and where they could always find a good gin rummy game. Yes, the work at times was demanding and tiring, but it was generally agreed that no one worked up a sweat. Why, then, did some decide to leave the shop and start another life? They could always come back to visit and get a free copy of the latest edition of the paper; but visiting, they knew, would be very different from entering the composing room to do a day's or night's work.

Most had been in printing their entire working lives. They started by sweeping floors or becoming proof boys. Then they served six-year apprenticeships and finally became journeymen printers. One described his life as a printer this way:

> I was sixteen when I started. I was big and strong. I swept the floor, cleaned the presses, and ran errands. I did everything I was asked to do for $5 a week. Then came my promotion. The foreman put a stick in my hand,[4] and with great patience taught me to set type by hand. Later I went to the union school for printing, and after serving my time I received my union working card.

Another stressed how happy he was to become a printer:

> I saw a sign, "Boy Wanted." I took the sign down and went upstairs to a small print shop.
>
> I was fourteen. I asked myself, What am I going to be? A street cleaner? A carpenter? A policeman? My brother-in-law was a printer, and I decided to try it. I ate it up, and I still do!

The work history of these printers approximates a career in that they identified with their occupation and had a succession of related jobs, all within the same industry.[5] But it was not an easy career. These men had gone through the Depression and had to struggle to stay in printing and to get their union card. "What can a man say who went through real hard times? I stuck to printing. I bounced around working as an operator, a makeup man, a proofreader. In 1937 I found my first steady job."

By 1976 men such as these had worked forty, fifty, and several over sixty years in the industry. Some were well into their seventies and even a few in their eighties, but they continued to work. Younger printers introduced us to them, and more than one pointed with pride to an older coworker: "He's eighty, and he's here every day." A few remarked that it was not all that easy to work next to an eighty-year-old, but they took it in stride. "Sure we carry some of the older men, but work is their life."

But not all of them, as the saying goes, "die with their aprons on." Postcards from retired workers describing the good life in Sun City, Arizona, and in Fort Lauderdale, Florida, are prominent on the bulletin board of the composing room and chapel office. Nonetheless, the popular view has been that printers work until poor health compels them to retire or, more graphically, until "they have to be carried out of here."

The longer a printer works, the greater his retirement benefits from the union.[6] In addition, these printers participate in a company pension plan provided by the newspaper that employs them. Exactly how much money a printer will receive as a retiree varies by age, years in the union and at the paper, and the retirement option he chooses. For example, he will receive smaller monthly payments if he elects survivor benefits or if he had been ill for a time while he was working. In 1976 the minimum was $110 per month for the printer who had been a member of Local 6 for twenty years. The average was about $200 a month, but it could be as high as $500 for the printer who was a foreman and who had been in the union forty years. In addition, every printer who retires at sixty-two years of age or older is eligible for social security. Counting everything—social security for his wife and retirement checks—a retiree could have an income of between $200 and $250 a week.

No single retirement incentive can be expected to appeal to all workers, let alone to printers who see retirement as a very personal decision and whose union contract stipulates that the decision is to be left entirely to the individual. Reportedly, during the contract negotiations management did not even raise the issue of a fixed age for retirement, or what is usually called mandatory retirement. The fact that the average age of the workforce was as high as fifty-six was probably a consideration,[7] but voluntary retirement suited management's interests as well. The publishers were not at all certain how many printers would be needed as they tooled up for automation. Maybe there would be an excess of men, and maybe not. Business could expand—at least one paper was thinking about taking on commercial work—and no one was certain how well the automated equipment would work under the pressure of several daily editions.

During negotiations the union pressed unsuccessfully for a reduction in the hours worked per week. An alternative that was agreed on by both union and management provides for a six-month paid productivity leave (PPL) to be taken during the life of the contract. The timing of the productivity leave and whether it is to be taken all at once or in smaller segments require management's approval. Any printer who retires, however, receives six months' pay in lieu of leave time (or for that portion of the leave time which has not been used).

As a result, the printer who retired during the first six months of the contract automatically received six months' pay plus the $2,500 bonus.[8] The bonus, as one negotiator put it, was to "sweeten the package." Together, the six months' pay and the bonus amounted to about $11,000. In sum, the contract provided incentives to encourage, but not force, retirement. And, in the words of both foremen and printers alike, neither management nor the union exerted any pressure on the composing room force to retire during these six months.[9]

Union's and Management's Views of the Incentive

The records of Local 6 show that 273 printers, or 13 percent of the work force at the three papers, retired during the first six months of the contract, the bonus period. About the same proportion retired from each of the three papers, thus ruling out the possibility that there was anything special about being an employee at the *Daily News, New York Post,* or *The New York Times* which influenced their decision.

In the union's view, the small number of men who took the bonus was no surprise. They say that the $2,500 provided little or no incentive because the men who took it were going to retire anyway. After all, printers at these papers knew in advance of voting on the new contract that it would probably include a retirement incentive, and those who had already decided to retire merely delayed the date to capitalize on the bonus. This is called "stacking up," and it is a common practice when contract negotiations are underway. It means that workers postpone any decision they can in anticipation of better benefits with a new contract. Since these printers knew as early as October 1972 that the union intended to negotiate a new contract, those thinking about retirement could do so at some leisure. Moreover, the old contract expired in March 1973, and the new one was not signed until July 1974 with the *Daily News* and *The New York Times* and not until September 1974 with the *New York Post*.

Historically, turnover has been low among printers. For example, sixteen is the average number of years printers have been employed at *The New York Times*. Moreover, attrition through retirement, disability, or death also has been low, about 5 percent, or ninety printers a year at the three papers.

At the time the contract was signed, some management officials thought that $2,500 would be an incentive, but now they share the union's view. A few may be rewriting history when they talk about the impact of the bonus, but they, too, think that even if the amount had been $1,000 more or less, the same number of printers would have retired.[10]

One requirement imposed on printers who took the $2,500 bonus was that they notify Local 6 in writing that they would not seek work as printers in the geographical jurisdiction of their Local. Those who may have thought about taking the bonus but who wanted to continue to work had two choices: leave the industry or leave New York City. Since most knew only printing—few had done other work for pay—it was difficult to imagine another kind of job, and relocating as a printer posed a big problem. Their current job was secure, the wage scale was good (average $350 a week), and where would they go? As one union official put it, "Who wants to go to Cornwall [New York] where the wage scale is much less? And, jobs for printers are scarce in big cities like Chicago and San Francisco."

We do not know how printers felt about the bonus at the time it was announced. But we have considerable information obtained after the bonus period from men who took it and from those who did not.

Those Who Took the Bonus and Those Who Did Not

To analyze the impact of the $2,500 bonus, we chose to examine two groups of printers who were alike in one important respect. They had passed their sixtieth birthday, and they had accumulated twenty-five years of employment credit as members of the ITU. The combination of age and years of work entitled them to full benefits from the ITU Negotiated Pension Plan. A total of 140 such printers took the bonus, and 209 declined it.

In the spring of 1976 we interviewed a sample of 130 of the 209 who decided to continue working and 87 who had retired.[11] We asked them about their work, the new contract, plans they had for the future, why they took or rejected the bonus, their attitudes toward retirement, and background information—how many years they had been working; whether they were day, night, or lobster men; their education; marital status; and the like.[12]

By no means can it be said that the eighty-seven were early retirees. Although at present more than one-half of the applicants for social security are under sixty-five, the average age of these printers at the time that they retired was sixty-eight. Three out of four were over sixty-five before they stopped working, and as many as one out of four was over seventy.

They told us that they live on about $10,000 a year which comes from pension checks, social security, and, for some 85 percent, interest from savings.[13] Considering expenses and income, three out of four say that they are living as well as or better than they did before retirement.[14]

Money in the bank is undoubtedly important, but those who are particularly likely to enjoy a higher standard of living are the 40 percent whose wives have worked or are still working. These women, usually nurses, school teachers, or secretaries, make a noticeable contribution to the incomes their husbands report. For example, among retirees reporting the highest income ($1,000 or more a month) three out of four have wives who are either working or retired.[15] In contrast, only one out of four who report the more modest income of less than $700 per month has a wife who is or who has been employed.

Almost to a man, these retirees have fully withdrawn from the labor force. Only two were doing any work for pay, and only six said that they were looking for work; but they are seeking part-time, not full-time, employment.[16] Work may have been central to their lives at one time, but now they opt for leisure.

As a group, they are vigorous and spirited. Most enjoy good health and have money in their pockets. To say that they have retired or that they have become senior citizens suggests that they are "over the hill." Although less so now than ten years ago, the very word *retirement* still has a negative ring. It signals that one is old and that he can answer the question What do you do? only in the past tense—"I was a printer." Many retired printers, however, have a different answer to this question. But first, how do they compare with the men who rejected the bonus?

Background Characteristics

In several respects the retirees and the working printers we compare are similar. They are second-generation citizens, typically of Irish, Italian, or Jewish descent. Some are German-, Scottish-, or English-American, and one describes himself as a "Heinz 57" variety. About four out of five report that one or both parents came from Europe; these are the men who call themselves Irish-Americans, Italian-Americans, and so on. Only a few identify themselves as simply American. One whose ancestors came here in 1635 called himself "American-American."

Many got their start in the trade with the help of a brother, a friend, an uncle, or their father. One out of two said that at least one other member of their family is or was a printer. "My father was a linotyper and a good one. Printing is in my blood."

The tradition of becoming a printer was especially strong among the Irish, with three out of four having come from a family of printers. Although a number were quick to say, "The Irish have a sharp eye for printing," their sons have not entered the trade. Nor have the sons of other printers gone into the industry. Less than 5 percent of all the printers and retirees, including the Irish, said that one of their sons has chosen printing as his career.

Four out of five are married. About 10 percent are widowed, and the remainder are divorced or separated or never married. Most had no more than a high school education, and this is especially the case for the retired man. More importantly, they are an excellent example of the upward mobility of the working-class man in the United States. They did not go to college, but a sizable majority gave their children that opportunity, whether they had one, two, or three children. And, it should be added for the record that the daughters of these men were as likely as their sons to have attended college.

The educational achievements of their children are a source of pride and satisfaction, and a number give all the credit to their union.

> In the early years, it was tough. I had only enough money to buy shoes for the kids *or* tires for the car. Then, it got better. Now I make a good living. I say, "Thank God I'm a member of Big Six." I even sent my kids to college!

As one would expect, most of these men are Democrats rather than Republicans. Even the 30 percent who label themselves Independent identify more closely with the Democratic than the Republican party. Both retired and working printers, especially the older men, go to church or synagogue services and value religion as an important part of their lives.[17] In sum, both the men who took the bonus and those who did not are second-generation, blue-collar Americans, strong union and family men whose children have become doctors, lawyers, teachers, and the like.

Attitudes toward Automation and Work

Before the retiree had stopped working, some sections of the composing room had already been automated. In others, the trial-and-error phase of introducing new equipment was underway, as well as the process of retraining. In addition, stories and rumors were everywhere. The new, complex machines were intriguing, but would operators be obsolete? What kind of skills would machinists need? Maybe there would not be enough work to go around, and the men would be standing around idle and bored. Or, work could turn out to be more interesting and challenging the new way.

The men told one another everything they saw and heard about the transition to an automated shop. They were so efficient about keeping their coworkers informed that word about a change at *The New York Times* could reach men at the *Daily News* before all the foremen at *The New York Times* had heard about it.

Almost every retiree was certain that once the composing room was fully automated, it would be cleaner and quieter. They accepted the fact that printers would no longer wear ink-stained aprons and green eyeshades while they sat on high stools or stood at long tables. The air would be cleaner, the noise gone, the accidents from molten lead a thing of the past, and their hands would not be black with grime. Printers' ink may have been the "sweetest smell on earth," but it also meant that there was dirt, clatter, and risk of injury from the hot metal.

In chapter 2 we saw that a majority of the printers accepted automation as inevitable and thought that it would make their own work physically less strenuous and as intellectually rewarding as what they had been doing. Both the retired and the group of working printers with whom we compare them were asked whether they thought automation would make the work of a printer more interesting, and their judgments were similar. Two out of five of those who took the bonus and those who did not saw automation in this light.

Only two retirees said that automation had dictated their decision to retire, and both expressed themselves exactly the same way. In large black letters they printed on the questionnaire "Automation!"

It may be surprising that only two retirees were so angered by automation that they blamed it for their retirement. This does not mean, however, that every other printer, working or retired, welcomed automation. Some plainly did not, and many others were watchful, even fearful, about how it would transform their work. Still, most men did not take or reject the bonus because their prestigious craft was being automated.

Moreover, the majority of men both who took the bonus and who did not said that they enjoyed working at the papers and that they liked their coworkers. In general, the bonus takers were more likely than those who declined it to have said that their work gave them a sense of satisfaction and that it was interesting. We attribute this to the nostalgia of retirement.

The group of men well into their seventies who continue to work is the one exception to this generalization. Virtually every one of them describes his job as really interesting; no other group feels so positive about work. This difference suggests opposing definitions of the meaning of work and of retirement.[18] One definition is that work is a livelihood, and some men have economic stakes in their job which propel them to work longer. The other is that work is a man's community and his life. Without the composing room, the excitement of getting the paper to bed, and the companionship of fellow workers, life has little to offer. After discussing the response of both retired and working printers to the idea of preparation for retirement, we turn directly to these two themes to help explain why some men took the bonus and others did not.

Preparation for Retirement

Preretirement seminars, easy-to-read summaries of social security benefits, advertisements by savings banks, television programs such as "The Retirement Revolution,"[19] and articles in newspapers and magazines are drawing more and more attention to the importance of preparation for retirement. The argument is that prospective retirees need practical information about legal and financial matters, about health and life-style considerations. None of these efforts is directed to telling a person when to retire or what to do in retirement, but the hope is that providing information and talking about the idea will reduce uncertainty and fear about making the decision and hence ease the transition.

Preretirement seminars are a recent development. Many are offered by employers, participation is voluntary, and the sessions are held during working hours. An employer may post a notice prominently in the employees' cafeteria: "What you always wanted to know about retirement (but were afraid to ask)—eight weekly two-hour sessions."[20] Another may insert a memorandum in the pay envelope of employees fifty-five and older.

Research indicates that employees benefit from such programs, but it should be remembered that only those who are interested in thinking about retirement tend to take advantage of them.

With few exceptions neither retirees nor our comparison group of working printers had the opportunity to attend any formal preretirement classes. They could, of course, do a number of things on their own, and we asked retirees and working printers who expected to retire by 1985 (88 percent of the 130 in our comparison group) several questions about preparing for retirement.

Overall, the number of retired and working printers who spoke with people about retirement, read a pamphlet or an article, or put aside some money is considerable.

	Men Who Took the Bonus ($N=87$)	*Men Who Did Not* ($N=114$)[21]
Read anything on retirement	53%	45%
Talked to other people[22]	51	77
Saved money	49	74

The bonus takers were less likely than the men who decided to continue working to have talked to other people about what retirement might be like. The changeover to automation in the composing room and the men's knowledge that management wants them to retire sooner rather than later may have made retirement a more frequent topic of conversation both at work and away from work than it ever was in the past.

The bonus takers also were less likely to have saved money for retirement. It is possible that these men started to think about it only in the last few years, given the tradition in the industry until recently of not retiring. If so, being late deciders, they were less able to have saved money specifically for retirement.

Printers also talk among themselves about retirement. Typically, younger men talk about it more than older ones, especially those who are thinking actively about it. Moreover, younger men talk about it with enthusiasm and make it sound like a long, paid holiday. They will be able to do everything they cannot do now; they will use their time as they wish, and they will have money in their pockets.

Somewhat poignantly, working printers said that the surest sign a coworker is about to retire is that he stops talking about it.[23] He may have approached the idea gradually—picked up a pamphlet at the union office and then put it on a shelf for a while. Or he may have moved much more quickly. This is the printer who takes a trip to Florida with his wife, sees a few friends, looks at condominiums, and almost before he realizes it, makes a down payment on one.[24]

One foreman who has seen many printers approach the last days of work described the experience this way: "He [the prospective retiree] gets fidgety as hell. He has been in the shop over forty years. He is leaving a home, and he is about to go on a fixed income. No wonder he is nervous!"

Although these days can be tense, sometimes another printer will take the time to talk with the prospective retiree about men who are enjoying retirement. "Remember Frank? He's got enough money. He travels. He doesn't complain about not working."

What these observations point up is that making the actual decision to retire is only part of the experience. The prospective retiree's need for social support from coworkers may be considerable between the time he files retirement papers and the day he actually stops working. It is also our sense that workers who see this taking place may well be the ones who gain most: they have the opportunity to rehearse, as it were, an experience they are likely to have one day.

No one knows for sure whether reading or talking about retirement was an incentive for some men to make the decision. We did not even find a relationship between looking forward to retirement and reading or talking about it among either the bonus takers or those who declined it. The comments volunteered by the men themselves suggest why those who did not look forward to retirement were as likely as those who did to have done these things. For example, several printers who were doubtful about how good an idea retirement was quoted a person or referred to an article which they had read that reinforced their negative attitude. Two described their experience this way: "Most people I talk to say it is the worst thing in the world." "I've read some bad stories about retirement."

On the other hand, the printer who looks forward to retirement finds people to talk to and things to see which support his thinking. "My union brothers [whom] I see are doing swell. Read the postcards on the bulletin board from all over the world, you'll see what I mean." Then there was the printer who said, "No matter what you think about retirement or no matter what someone says about it, you can find someone who says the opposite."

These remarks suggest that the response of printers to preparation for retirement, as measured by reading or talking about it, is another instance of selective attention, a concept which public opinion surveys have shown to be widespread. To cite only one example, in election campaigns people listen to the speeches and read the materials of candidates whom they support, not of those whom they oppose.[25]

The Decision to Work or Retire

Most printers with whom we compare the bonus takers expect to retire by 1985, the date that the contract expires. This target date is undoubtedly too distant to suit management, because the composing rooms at the papers will be fully automated before then.

Some are uneasy because they worry about what is going to happen. How long will it take for automation to wipe out their jobs? Then what will they do? They know that management is not going to want to pay them to come to work and just sit around.

Most, however, see this as management's problem. Since the contract guarantees them job security for life, they are free to decide when they think that it would be to their advantage to retire. They also know that the contract provides six months' pay in lieu of the productivity leave whenever they retire. And, with the cost-of-living clause built into the contract, the dollar value of the productivity leave increases with time. But there was a more important reason for them to pass up the $2,500 bonus.

In 1976 the average age of the printer who declined the bonus was sixty-five, and that of the men who accepted it was sixty-eight. But this figure blurs an important distinction: Was the man under or over sixty-five during the six-month bonus period? As may be seen below, 62 percent of the men who did not retire during the bonus period were under sixty-five at that time, but this was the case for only 24 percent of the men who did retire. Moreover, the printer not yet sixty-five who let the bonus go by was closer in age to sixty-two than to sixty-five; the opposite pattern holds for the man under sixty-five who took the bonus.

Age at the Time of the Bonus	*Men Who Took the Bonus* (N=87)	*Men Who Did Not* (N=130)
Under 62	5%	23%
62-64	19	39
65-69	47	23
70 or older	29	15

The distinction, under or over sixty-five, is critical for understanding the response of these two groups of men to the $2,500 incentive to retire.

Men under Sixty-Five Who Declined the Bonus

Two economic facts were uppermost in the minds of the men under sixty-five who rejected the bonus. First is eligibility for maximum social security retirement benefits. Printers, as most other workers, do not know exactly how much money they will receive in retirement benefits from the company and from the union until they file for retirement; but from their comments it seems that every working and retired printer knows the eligibility rules for social security. A worker may retire as early as sixty-two and receive social security, but the amount of his monthly check will be permanently less than that paid to the worker who delays retirement until he reaches sixty-five. The total benefits over the years will be about the same as those to the worker who waited until sixty-five, but the younger retiree receives smaller installments to take into account the longer period for which it is expected that he will receive social security. The amount of the reduction depends on the number of months for which the individual receives checks before reaching sixty-five. Thus the person who retires at sixty-two receives smaller monthly checks than the one who waits until sixty-four. In addition, the wife or dependent husband must be at least sixty-two to start collecting social security along with the retired worker.

There was a second economic reason for the men under sixty-five to continue working. If they took the bonus, they sacrificed the union package of health benefits with the exception of Blue Cross coverage for themselves, their wives, and dependent children provided they had accumulated twenty years of service (or work) credit in New York City. Doctors' fees, dental expenses, and eyeglasses, for example, which are part of the working printer's benefit package were, however, the retired man's full responsibility. If his wife was working and she had health insurance through her employer, he had some protection; otherwise, he had to fend for himself. Further, no one who applies for social security retirement benefits is eligible for Medicare, the federal health insurance program, until he reaches the age of sixty-five. A sixty-four-year-old printer summed up the predicament that he and his coworkers would have confronted had they taken the bonus: "It is a whole year before my wife is sixty-two and can collect hers. Just one illness could wipe out our nest egg."[26]

Most private pension plans, including the ITU Negotiated Pension, increase in value, the longer a person works. The man who has worked steadily for thirty years, for example, receives a higher monthly retirement check than the one who retires after twenty-five years of work. What the printer under sixty-five had in mind, however, when he rejected the bonus was not so much the negotiated pension as the desire for top social security dollars and health insurance.

A majority look forward to retirement, and four out of five expect to stop working at sixty-five. Those who plan to work a year or two beyond sixty-five are delaying their decision either because they are married to women who will not be eligible for social security by the time that they themselves are sixty-five or because they have a youngster in college, a mortgage to pay off, or some other kind of personal debt.[27] One put the blame on his wife: "The old lady hit me for a modernized kitchen. Now I owe money to the credit union, and I'll have to work a little longer." If these younger men retire within the next few years, and the data indicate that they will, the old-time printer who never retired will have become a figure of the past.

The main reasons these men decided to continue working at the time the bonus was offered underscore the importance of economic considerations. The most typical response was "I'll wait two more years and get social security and Medicare. Retiring sooner makes no sense to me." One out of two printers felt this way. Although other reasons were given for continuing to work, they were by no means as frequent:

Reasons Men under 65 Gave for Declining the Bonus[28]

Waiting for social security and Medicare	49%
Like to work	25
$2,500 was no inducement	21
Need money (building a house, son in college)	18
Personal (not ready psychologically, wife working, would be bored)	13
	($N=76$)

Even the men who said that they continued to work because they preferred to do not see themselves in the composing room after they reach sixty-five. In one way or another, they expressed the sentiment of this man: "I'll have worked long enough by then!"

That so few men under sixty-five—only 21 percent—mentioned the bonus when they explained why they were still working indicates what limited interest it had for them. "That $2,500 never fazed me. What would be left after taxes?" "$2,500 is peanuts. If they dangled a large sum in front of my eyes, maybe I'd reconsider." It is conceivable that a much larger sum would have "bought out" this man, but for most the amount of the bonus (and, for that matter, even its availability) was at best a secondary consideration.

Early Retirees: Men under Sixty-Five Who Took the Bonus

Almost all the early retirees were just under sixty-five, which meant that when they stopped working, they could begin to collect close to the maximum in social security and in a short time would be eligible for Medicare. Age was clearly in their favor, and they could retire with more financial security than a sixty-one- or sixty-two-year-old man. Further, for one out of three retirement really was not by choice. Poor health was the primary reason, and some of them were eligible for social security disability rather than retirement benefits. "It was getting hit by a truck that decided me.

With a cane you don't get around machines too good." "The arthritis in my legs got too bad to work." "I had a second heart attack, and on the advice of two doctors, I retired."

These men had liked their work, but poor health or a disability had forced them to retire. They often referred to the difficulties they were learning to live with, from limitations on what they could do and what they could eat, to the high cost of medical care.

As a group, they discounted the bonus as a factor in their decision. At first, their negative attitude seemed surprising, because retirees with health problems were most likely to have said that, considering income and expenses, they do not live as well as they did before retirement. Their financial position, then, suggested that the retirement incentive monies should have been important to them.

The reason why the money was not important can be explained more simply. Poor health had compelled them to stop working. With or without the bonus, their working days were over, and the bonus had little to do with their decision. The outlook of these men toward the bonus was also typical of the man over sixty-five who retired in poor health.

The number of printers in good health who were not yet sixty-five at retirement is small relative to the number sixty-five and older. All the same, they are a critical group, because they represent the new national trend toward voluntary early retirement. As recently as 1966 only 38 percent of the workers applying for social security were not yet sixty-five, but by 1976 the figure had jumped to 66 percent.[29] The experience and attitudes of printers who opted for early retirement provide insights into this dramatic change in retirement patterns.

Further, although the national norm has become to retire before sixty-five, these men have worked all their lives in a trade which has traditionally dismissed the idea of retirement. Thus, these men are among not only the first to retire voluntarily, but also the first to choose relatively early retirement.

The main conclusion of a study of automotive workers is that the decision to retire (or plans to retire) before sixty-five depends on financial circumstances, in particular, expected retirement income (Barfield and Morgan 1970). Just as with the automotive workers, anticipated retirement income was clearly important in the decision of printers who retired before sixty-five.

A larger and more recent study of five years in the working lives of middle-aged men in diverse occupations suggests that it is not only "bread and butter" factors which dictate whether a man retires before sixty-five, but also commitment to work.[30] If we measure the printer's commitment to work by one of the two indicators used in this cross-sectional study (namely, doing some work for pay even though one's retirement income makes it possible to live comfortably), the retired printers have a weak commitment to work. Their motto, as one expressed it, is "Once I'm out, I'm out." In this respect they personify the economic man. Work was a livelihood and a relatively good one, so much so that they could send their children to college; but now they have no desire to work.

By no means, however, did they retire alienated from their union or from work. They credited their union with having made retirement possible, and they valued the craft which they had mastered. "Even the frustrations [at work] were in a way enjoyable," one remarked. Some had grown tired of the conditions surrounding work—traveling late at night, working odd hours for so many years, no longer having the freedom to hire a substitute when they wanted to take time off—but they are proud to have been printers. They liked the pace and excitement of composing the daily paper and having been among the men who had the "first look" at tomorrow's headlines.[31]

In a very real sense, these men entered retirement with the odds in their favor. They had liked their work; they voluntarily decided to retire; and as one would expect, their attitude toward retirement was positive. All but one had looked forward to it, and this one said that although he had some reservations about the idea, he had decided to accede to his wife's desire to move south.

In sum, the differences between the men under sixty-five who took the bonus and those who did not relate, first, to a gap in age of a few years and the financial considerations this implies and, second, to health. And even though work had been a meaningful part of their lives, they decided to give it up.

Men Sixty-Five and Older Who Declined the Bonus

Some printers talked about the trauma of turning sixty-five just as marking one's fiftieth birthday is said to give a woman pause and the fortieth to be unnerving for a man. The printers who feel that sixty-five separates the middle-aged from the older man are undoubtedly thinking about the fact that sixty-five is the age when one is fully eligible for social security. What follows, then, from this reasoning is that the person who takes social security admits to being old. The prototype of this outlook on life is the printer who has reached sixty-five but who does not want to retire: "When you're sixty-five and you retire, people say, 'Oh he's an old man.' "

Men such as this printer have two advantages, one which we refer to throughout this book, namely lifetime job security, and a second advantage which is of equal importance, good health. Some 84 percent of those sixty-five and older who have continued to work judge their health as good or very good. They emphasized how healthy they felt by saying, "I'm in excellent shape" or "My health is topnotch." A few have high blood pressure, arthritis, or an angina condition, but none of these problems prevents them from getting to work. At least one, a seventy-three-year-old, attributes his good health to the fact that he does work: "I'm on my feet when I work. This makes the blood circulate better. After a day's work, I sleep better at night." Others, as we shall see, want to continue working for different reasons.

A number of studies have shown that aversion to retirement increases with age and especially in the sixties when retirement is about to take place.[32] We see a similar pattern among printers. While three out of five of the group under sixty-five look forward to retirement, only 37 percent of those between sixty-five and sixty-nine and fewer still (21 percent) of those seventy and older have this attitude. Most of those sixty-five and older who said that they wanted to retire are still in their mid-sixties, and they expect to stop working within a year or two. Like some of those under sixty-five, they are working a short while longer to make the last payments of a mortgage, or waiting for their wife to reach sixty-two, or temporarily contributing to the support of a son or daughter. The situation, however, is very different for the majority of those over sixty-five who either have mixed feelings about retirement or dislike the idea.

Two out of five of those over sixty-five are not at all sure that retirement is in their best interests. These men are considering it, but from their comments it is clear that they are not ready to "jump into retirement," as one described what it means to him to retire. "A part of me wants to retire, and a part of me does not." "I am trying to adjust my mental set to it, but I don't know if I can." "Frankly, I'm scared. What will I do with myself?"

About one in three dislikes the idea and did not hesitate to say so: "I hate the thought." "I think straight and fast on my own. Retirement is not for me." Among these men are some who flatly stated that they will never retire. They have no interest in it, and their negative attitude is well known to fellow workers. Retirement to them means the end of living: "If I stopped working, I'd die." "What am I, crazy? I wouldn't know what to do with myself. I know men who started going to bars all the time when they quit working." "I'd be lost if I were not working. Work is therapy."

Men seventy years of age and older are particularly likely to value the opportunity to continue working. Recall that almost everyone in this age group said that his job was really interesting, and most express no regret that automation is changing the nature of their work. "Automation makes no difference to me. I'll adjust. It just means using another tool."

Do they get satisfaction from work? Yes, but not so much from the kind of work they do, but because work gives them a place to go and something to do. "There is not much to do away from here. What am I going to do, watch TV every day?" "I'd rather work than hang around."

Further, one out of four was widowed, divorced, or single. Not one of these men looked forward to retirement, and more than one-half did not hesitate to say that they disliked the idea. Almost all of them have worked at the same paper and on the same shift for many years. In effect, the composing room has been their home, and they want to keep it that way.

> I consider retirement a mistake. Being in the composing room with these guys keeps me alive.
>
> I live in a hotel uptown. What am I going to do retired?
>
> It's not the money. I'm very much alone. I simply don't want to retire.
>
> For the last five years, I've been saying next year, but I am kidding myself.

Some of the older married men are also reluctant to retire, and they mentioned their wives when they explained their decision to continue working.[33] "My wife likes people around her. I'm an introvert. It is better for both of us if I work." "I'd never retire until my wife is eligible for social security. I wouldn't feel like a man retiring and letting her work."

To be sure, not every reason for continuing to work applies to every man, but one does not have to meet every criterion to decide that retirement is not in his best interests. Moreover, the arguments and explanations which these men offered make clear that they value highly their right to continue working. Only one thinks that a fixed age for retirement is the answer: "Sometimes I think mandatory retirement is easier. It would save me from wrestling with the decision."

Thus, barring a serious health problem or an unexpected turn of events, these men who do not want to retire will do their best to continue working.

How Important Was the Bonus?

Neither the union nor management, as noted earlier, thought that the $2,500 bonus changed any printer's mind about retiring. When we asked retirees how important the bonus was in their decision to retire when they did, 42 percent said it was very important, 37 percent attached some importance to it, and 21 percent said that it was of little or no importance. Although only five said the bonus was *the* reason that they retired when they did, several had good things to say about it. "It gave me a little more security." "I waited for the right opportunity [the bonus]."

The men under sixty-five and those seventy or older thought somewhat more of the bonus than those who had retired between the ages of sixty-five and sixty-nine. The younger men, especially those in good health, were eager to retire. Close to sixty-five, they were willing to take social security and to gamble that they would not have large medical expenses before they became eligible for Medicare. The bonus, plus six months pay, provided a financial cushion which made the incentive important in their decision.

The older retirees were also enthusiastic about the bonus, especially those married to women about their age. These men and their wives were in good health, and they probably reasoned that a dollar incentive such as this one might not come along again in their lifetime, at least while their health was good and they could enjoy it. The prototypical case was the seventy-eight-year-old retired linotyper who decided to stop working after sixty-two years in the industry because he wanted to do some traveling with his seventy-seven-year-old wife. He said, "I grabbed the bonus!"

The bonus did not have this kind of appeal to any of the printers who preferred working to retiring. Whether they were under or over sixty-five, $2,500 was a trivial amount of money. They all knew that they could collect six months' pay in lieu of the productivity leave whenever they chose to retire during the life of the contract. And given the automatic cost-of-living increases, the longer they worked, the more the productivity leave was worth. No wonder, then, that the attitude of these men toward the $2,500 bonus was "I never gave it a thought. Who are they kidding?"

A Successor to the $2,500 Bonus

By the fall of 1977 the pace of automation had quickened, and management at *The New York Times* decided that the men downstairs, as they referred to the printers in the composing room, were not retiring fast enough. They expected that in a few months there would be an excess of men, and they designed an incentive plan for the first fifty who would take it and voluntarily terminate their employment.

The plan provided $5,000 to every printer who stopped working at the paper, regardless of age, and for those not yet sixty, a formula for a larger retirement incentive. For example, a fifty-three-year-old would receive a lump sum of $17,000, and a forty-year-old would be entitled to about $30,000. The monies would be subject to normal withholding tax requirements, plus union dues, which totaled 4.5 percent of wages.

One printer described the response to the offer:

> At first many of us were excited about it, but when you thought about it, you realized it was an insult.
>
> How much money would you take home? I make $20,000 a year. I really only know my trade. I have twenty years to go to reach sixty-five.

> You do a lot of thinking before cutting the cord. Jobs are not easy to get. How can anyone under sixty be interested in the offer?

The net result was that nine men, not fifty, as the company wanted, took the offer. One who took it was about to go on disability retirement, and all the others were at least sixty years of age.

In retrospect, management thinks that they bungled the offer, but more in the way that they announced it than in what they offered. They simply posted a notice in the composing room and then sat back, fully expecting that many more than fifty printers would rush to take advantage of it. They added that they called union headquarters in advance, but, as expected, the union leadership did not take a position on the offer.

Management's view of the size of the offer and of its potential for encouraging a printer to stop working at *The New York Times* is very different from that of the printer. Management thinks the men did not see the opportunity which this incentive represented. They know well that a man who terminates cannot seek work as a printer in the jurisdiction of Local 6, but they believe that a man can be encouraged to seek a new career or move away from New York.

The example given by management was that a man with $2,000 can set up a small grocery store in his neighborhood and be supported if only seventy families patronize the store. Reportedly, one representative of management said that $5,000 could be used as the downpayment on a house, and a man could then relocate. The printers' retort is "I'd say you'd be lucky to have the downpayment for a Toyota after taxes."

At this point it would seem that management and the printer are far apart on the subject of an incentive to early retirement. This does not mean, however, that printers are not concerned about having a job to do. They do not want to come to work and just "hang around," as several expressed it. They want to continue working and to be productive. Moreover, they want their employer to make "barrels and barrels of money," as one said, because the more the paper grows and increases its readership and advertisements, the better it is for them. But if they agree to leave the industry and to begin a second career, it has to be something far more appealing than operating a small corner grocery store.

Summary

The $2,500 retirement incentive was anything but a "spur to retirement," as it was described in the press the morning after the contract was signed. It is not that workers who have the protection of voluntary retirement will never stop working and therefore have no interest in such a dollar incentive. Rather, the question is: When is it in my best interests to retire? The answer depends not on the offer of a bonus, but on one's definition of retirement.

To printers, retirement means withdrawal from the labor force, not the beginning of a second career. They know how tight the labor market is for the middle-aged man looking for work, and the market is particularly scarce for union printers. They could move outside the jurisdiction of their Local to look for another job, but if they found one, the pay scale would be lower. Some could become salesmen or go into real estate, but what would their income be and what job guarantee would they have?

As printers at the newspapers, they enjoy a good salary and job security. They also like the composing room and the people with whom they work. Most accept the automation of their craft as inevitable and are coming to terms with it. Further, their loyalty and respect for the union which they built reinforce their sense of worth as workingmen.

Most will retire, but the eligibility rules for social security and Medicare set the time frame for their decision. That most men who took the bonus were over sixty-five underscores how important the monthly social security check is for them. Retirees liked the gift of the bonus, but it was no more than that.

Finally, a small minority of older men—well into their seventies and a few even older—cannot bring themselves to retire. The composing room is their home, their community, their life. It is these men, above all, that the concept of voluntary retirement protects. Their coworkers "carry them," and their presence adds to the dollar costs of production, but only in the short run, because it is widely thought that this is the last generation of such men.

Notes

1. A number of studies (for example, Blauner 1964; Goldthorpe et al. 1968; Sheppard and Herrick 1972) examine the attitudes and behavior of workers, but not in the context of job security and voluntary retirement. Neither of these conditions necessarily guarantees job satisfaction, but the opportunity to study the meaning of work and of retirement under these circumstances is unusual.

2. The eligibility period for the bonus at the *Daily News* and at *The New York Times* extended from July 28, 1974, to January 28, 1975, and at the *New York Post* from September 8, 1974, to March 8, 1975.

3. *The New York Times*, July 29, 1974, p. 12.

4. A typesetting tool with a built-in thumb clamp.

5. For a discussion of some of the consequences of an orderly work history, see Wilensky (1961).

6. Technically, his major retirement benefit is not a union pension. It is an industrial pension (also called the negotiated pension) which is based on employer-employee contributions.

7. The 1975 collective-bargaining agreement for the commercial branch of the printing industry, known as book and job, stipulates mandatory retirement at the age of sixty-eight. The average age of the union printer employed in a book-and-job shop was forty-nine at the time their contract was signed.

8. It should be remembered, however, that the six months' salary is not a retirement incentive because it represents a paid productivity (or sabbatical) leave to which every printer is entitled during the life of the contract.

9. See Davis (1973) and Fields (1970) for a discussion of early retirement provisions as a mechanism to pressure people to retire.

10. Between 1977 and 1979 the management of each of the papers offered larger financial incentives to retire in an effort to reduce the size of the workforce. See the last section of this chapter for a discussion of the 1977 offer by *The New York Times* and chapter 7 for incentives offered by all three papers in 1978 and 1979.

11. For an account of the methods of selecting the sample, see appendix A.

12. See appendix B for responses to each item in the questionnaire.

13. The importance of expected retirement income has been well documented by previous research. See, for example, Atchley (1976) and Barfield and Morgan (1970).

14. At the time that they retired, these men were earning about $15,000 a year. With social security and income from their union and company pensions, their spendable income was not substantially less, and for some it may have been even higher than before they retired.

15. Presumably, retired wives are also receiving pensions.

16. Data from the Retirement History Study of the Social Security Administration suggest a similar pattern nationwide with respect to withdrawal from the workforce. Only 2 percent of the men retired in 1969 returned to work in the succeeding years (Bond 1976).

17. See Maves (1960) for an informative discussion and a review of the literature on aging and religion.

18. These diverging orientations toward work parallel those of the instrumental and traditional printer developed in earlier chapters.

19. Presented in the fall of 1977 on a national television network.

20. This announcement was posted at one of the papers included in the study.

21. This question was asked only of men who plan to retire by 1985, the expiration date of the contract.

22. The question to retirees was: Before you retired, did you talk to retired people about what retirement might be like? The question to nonretirees was: Have you talked about retirement with friends or relatives?

23. Some prospective retirees, of course, do talk about it. More often than not, these men boast about having made the decision. If a printer is not listening closely, he will think his fellow worker is already retired and living a life of supreme fulfillment.

24. We exclude from this discussion the printer in poor health. Typically, he is on disability leave and retires without returning to the composing room.

25. See Lazarsfeld, Berelson, and Gaudet (1968, chapter 9).

26. See chapter 6 for a discussion of recent improvements in the health and welfare benefits for members of Local 6.

27. A few of those under sixty-five (less than 10 percent) either do not plan to retire or are reluctant to even think about the idea. We discuss their attitudes when we consider the larger group of men over sixty-five who have no interest in retirement.

28. Percentages add to more than 100 because a printer could give more than one reason.

29. U.S. Department of Health, Education, and Welfare, as reported in *U.S. News and World Report* 86, no. 13 (April 13, 1978):50.

30. Commitment to work was measured by an index composed of responses to two items: whether a man would continue to work if he had enough money to live adequately, and whether he would seek work if permanently laid off. See Parnes et al. (1974, volume 4, chapter 5).

31. Not all retirees leave their jobs with the satisfaction and positive orientation of printers. Studies of teachers, clerical workers, and factory workers, for example, show that a majority worked essentially to earn a living and derived little satisfaction from work (Atchley 1971, 1976).

32. For a discussion of the literature on attitudes toward retirement see, for example, Donahue, Orbach, and Pollak (1960); Riley and Foner (1968, chapter 18); Lehr and Dreher (1969); and Atchley (1976).

33. In chapter 5 we see the same kind of reasoning on the part of some of the retired printers who credit (or blame) their wives for the role they played in the retirement decision.

Appendix 4A

Table 4A-1
Comparison of Men Who Took the Bonus and Men Who Did Not
(percent)

	Did (N=87)	*Did Not* (N=130)
In printing entire working life	83	82
In poor health	12	2
Married	79	80
High school diploma	91	78
Thought automation would make work more interesting	38	41
Said work gave them a feeling of a job well done	95	78
Regarded most of the people as nice to work with	85	88
Found work really interesting	80	60

Table 4A-2
Attitude toward Retirement by Age

	Percentage of Men Who Took the Bonus		*Percentage of Men Who Did Not*	
	Under 65	*65 or Older*	*Under 65*	*65 or Older*
Look(ed) forward to retirement	76 (21)	50 (66)	61 (81)	31 (49)

Table 4A-3
Reasons Men Sixty-Five or Older Gave for Declining the Bonus
(percent)

Like to work	57[a]
Would be bored if not working	24
Need money (completing mortgage payments, building retirement home)	23
$2,500 was no inducement	18
Pension considerations, self or wife	16
Personal (supporting handicapped daughter, recently widowed)	14
	(49)

[a]Percentages add to more than 100 because a printer could give more than one reason.

5 The Retired Printer

During the past twenty-five years, retirement has been transformed from the privileged status of the wealthy to an established institution in industrial societies. In the United States, for example, the change has been so dramatic that 60 percent of the men between the ages of sixty-five and sixty-nine were still working full time in 1960. In 1972, the figure was only 21 percent (Sheppard 1976).

In general, the earliest to retire are blue-collar workers, then clerical, and somewhat later, professionals (Jaffe 1972; Parnes et al. 1974; Sheppard 1976; Bond 1976). The most frequent explanation given is that high-status-job holders derive more satisfaction and financial reward from work than those in lesser positions. Traditionally, printers were expected to work indefinitely unless forced out by poor health; and, indeed, the average age of the printers in our study at retirement was sixty-eight—hardly typical of the blue-collar worker's tendency to retire by sixty-five. The very fact, however, that so many printers are retiring is further evidence of the older worker's withdrawal from the labor force.

By 1976 the printers who took advantage of the $2,500 bonus and six months' pay had been retired about two years. Most were in good health and able to meet their day-to-day expenses, two essential considerations for a successful retirement. Some have found retirement exciting and rewarding; others have found it difficult and somewhat disappointing. As many as one out of five has not adjusted two years after leaving the job. This chapter describes and accounts for the different appraisals of retirement by printers.

Reasons for Retirement

The most important reason for the increase in the number of retired persons is social security and private pension plans (Rones 1978). Health also may be an important factor in the decision; but in recent years it is lessening, partly because work is not as physically strenuous as it used to be and partly because serious illness and disability affect people at a later stage in life than formerly (Palmore 1964; Pollman 1971; Bond 1976; Atchley 1976). The inability to find a job, various work rules, job dissatisfaction, the influence of family and friends, and attitudes toward leisure also play a part in the decision to retire.[1]

The reasons printers gave for retiring are very much like those of other workers who have withdrawn from the labor force.[2] They include health, a preference for leisure, the influence of their wives, and being financially able to stop working. Not all reasons, of course, apply to every man.

Reasons for Retiring

There were things I was looking forward to doing in retirement.	50%[3]
I felt I had enough money to retire.	38
I was tired of working.	35
My wife wanted me to retire.	29
I wasn't feeling well enough to work.	28
My children wanted me to retire.	18
My wife's health was poor.	13
Other (moved South, automation).	5
	($N = 84$)[4]

Most printers made the decision to retire essentially on their own. Only about one out of four gave as a reason "My wife wanted me to retire," and if their wife was the stimulus, their children reinforced it.[5] More generally, their answers indicate that some made the decision for negative or more neutral reasons while others did so for quite positive reasons, for example, "I was tired of work" versus "There were things I was looking forward to doing in retirement."

To better understand the different orientations of printers toward retirement, we have employed a distinction which has long been used to sort out responses to the question why?[6] This distinction separates the "pushes" from the "pulls" affecting decisions and allows us to divide retirees into three groups, depending on the explanations they gave as to why they retired.[7] We have termed these three types of men the retreatist, the activist, and the involuntary retiree.[8]

Retreatists (thirty-six printers) have accepted retirement, but they have an essentially passive, reactive outlook toward it. They explained their decision by saying such things as

> After fifty years of working and varicose veins, I had had it.
>
> The neighborhood was getting worse, muggings and holdups all along Eighth Avenue.
>
> Forty-five years of work—enough!

I just wanted to take it easy.

The Mrs. kept after me to retire.

I had enough money.

These men drifted into their new role rather than taking it on with enthusiasm. At best they expressed mild interest in retirement. Their financial situation allowed them to retire, their wives and children told them to retire, or they no longer wanted to work. They gave little or no indication of thinking about the decision from the perspective of what they would do in the future. They were seeking security or "peace of mind," as one phrased it, rather than thinking about another stage of life and actively substituting leisure for work. Even the few who mentioned the bonus as a contributing factor saw it as security rather than as an impetus to a new life-style. In other words, retirement was not pulling them, they were feeling pushed.

In contrast, activists (twenty-seven printers) explained their decision in terms of the opportunity retirement offered, stressing the pulls rather than the pushes, or family pressure to make the decision. They appeared confident, enthusiastic, and eager. Unlike retreatists who retired to take it easy or to put an end to their wives' complaints, activists saw retirement as a chance to begin a new phase of life:

I want to travel.

I have academic goals.

I want to give my son a hand with his business.

I love sports.

I play a musical instrument. I'm a bug on taping and recording.

The Involuntary Retiree

There is one remaining group, the involuntary retiree (twenty-one printers) who felt neither pushed to stop working nor pulled by the prospects of retiring. Nonetheless, they are worthy of description because they represent an important type of man. Their reason for retirement was poor health, either their own or their wife's. They had a heart condition, severe hypertension, arthritis, or another chronic condition. They had taken their physician's advice to retire, or they had to care for a wife whose health was failing: "Ill health—high blood pressure and sight deficiency." "It was emphysema. I had to stop working." "For the sake of my wife's health, we moved to Florida."

Poor health sets the tone of retirement for these men. They are able to care for themselves and to maintain their home, but health is a pressing concern. Moreover, unlike retired automotive workers who report that their health has improved with retirement, this has not been true of these men.[9] A total of 40 percent, in fact, have been hospitalized during the past twelve months.

Poor health, of course, brings financial problems. One-half said that, considering income and expenses, they are not living as well as before retirement, and one out of three has less than enough money to live comfortably.

Most are over sixty-five and have Medicare. But Medicare hardly covers all medical expenses, and certainly not the expenses of a special diet or hiring someone to help with housekeeping chores. Some are quite bitter about their financial circumstances and blame their union. They complained that "after all those years of working and paying high union dues," their pension benefits have been reduced and they do not have health coverage to help meet the cost of eyeglasses, medications, or dental visits, all of which working printers enjoy. They resent the fact that these benefits are no longer theirs, just when they need them most.

All the same, they, more than retreatists or activists, reminisced about the life they enjoyed so many years as union men and skilled craftsmen. Their comments about the past take on poignancy against the backdrop of their present difficulties.

> In a union you will find strength. You derive a better living wage, sociability, and good working conditions. All these I found at Typographical Union No. 6. No better union for the working man. I never regretted being a union man.

> What have I done? I have been an errand boy, pressroom helper, press feeder, a pressman, a salesman during the slack season, an apprentice, a printer, a linotype operator, stonehand, ad man, and proofreader. I was also a paper cutter, but not at the newspaper, and for a short time I had my own shop.

> I love my fellow workmen. Loved to set type on *my* linotype. I loved the *Daily News*. I always tried to do a good job and give a good night's work. I know my work was appreciated.

Despite poor health and financial problems, involuntary retirees are as likely as the other two types of printers to visit members of their family, play a game of cards or chess, or go to a meeting of a senior citizens' club or the Knights of Pythias. Yet these kinds of things seem not to be enough, probably because they take up only a small part of the day or they happen less often than the men would like.

As a group, these men appraise retirement as a trying period in their lives. Of all the retirees, they were least likely to have looked forward to retirement, and two years later as many as two out of five had still not adjusted.

In sum, the misfortune of poor health is their chief problem, and it constrains what they do and how they live. The literature indicates that poor health and financial worries are the two major factors which predict a difficult adjustment to retirement (Shanas 1968). The experience of these men suggests that these two factors cannot be separated. Financial concerns go hand in hand with poor health, especially for the blue-collar man who has limited financial resources and minimal health coverage.

The Retreatist versus the Activist

Unlike the involuntary retiree, both the retreatist and the activist were in good health, but it takes more than good health to ensure a smooth and easy transition from work to retirement. One's orientation toward the new role is also important.

The activist saw retirement as an opportunity, and he was enthusiastic and confident about his new life. He had enjoyed being a printer, but those days were over. Now he was absorbed in making his life as a retired person as full as possible, and he had little interest in retelling stories, good or bad, about his working days.

The retreatist, in contrast, made a point of describing how weary he had become of the conditions surrounding work. He was proud to have been a printer, but now two years after retirement he talked about the things he did not like about work, especially if he had worked odd hours. As though it were yesterday, he recalled how tired he had become of riding the subway in the middle of the night or of trying to find a safe place to park his car. "Going home at 1:15 a.m. was nerve-wracking." "I felt my life was in danger working in a slum area. I was afraid of being accosted by hoodlums." "The hours were terrible."

Twice as many had worked the lobster shift, and it had gotten to them. When they were younger, this shift had some appeal, but not with advancing years.

> You had a free life. You didn't have to rush home for dinner. During the day you could charter a boat and go fishing or take an extra job.

> There was no heavy brass on the floor, just the shift foremen. And, you knew everyone—reporters, photographers, the editorial staff. You could take an extra ten minutes for lunch, have coffee when you liked, hot bagels on Saturday night. Then, there was the 5 a.m. poker game.

They had learned to "sleep on a picket fence," as one put it, but by the time they neared retirement, they had grown tired of working when their family and friends were sleeping or going out for the evening. One expressed his feeling very simply: "You get tired of having soup when your wife is having breakfast."

The fact that the new contract eliminated the substitute system was also distressing, because they liked the good old days of being able to hire a substitute so that they could work a shorter week or take an extra vacation. "No more substitutes meant asking the foreman for time off. Not me!" The substitute system had permitted them to taper their workweek and in effect to retire gradually by working fewer days a week. This is what they preferred, and the idea of returning to a full workweek was so unpleasant that they had no choice but to retire.

Thus, it was not work itself but the conditions which went along with it (odd hours, a new regulation, and traveling to and from work) that displeased them. At one time these conditions had been acceptable, even attractive. Most had been assigned to the shift which they preferred, and they were glad to have had steady work; but now their attitude was "I'd had it working!"

Talking about the past, rather than involvement in the present, is not the only distinction between retreatists and activists. Activists, more than retreatists, had completed high school or spent some time in college. This may not seem like much of a difference in educational achievement given today's standards, but when these men were young, most people quit school at fourteen or fifteen years of age. The activists, however, made a point of going to school after their day's work until the Depression hit them. Then they had to look for a job, they had no time for school, and printing became their life's work.

Both activists and retreatists had married young and raised children, but activists were more likely to have married women who had either gone to college or at least completed high school. In turn, their wives were more likely to have worked, and the children were more apt to have gone to college.

One additional distinction should be singled out, namely, involvement with the union. To be sure, both types of men regularly voted in union elections, abided by the work rules, and willingly helped a union brother in need. But it was the activists who sought leadership roles in the union and participated in union affairs. More than retreatists, they had been chapel officers or candidates for office in Local 6 or another ITU local and had attended social functions of the union.

Thus, we know that retreatists and activists differed not only in their reasons for retiring, but also in their background and accomplishments. In brief, retreatists had limited horizons and aspirations in contrast to the activists, who were men of higher achievement, more integrated into their union, and more socially mobile. What were their respective attitudes toward retirement and their experiences in their new role?

Attitudes toward Retirement

Since retreatists were tired of working, ready to slow down, and financially able to retire, one might think that they would be most likely to have looked forward to retirement. This is not so. About one-half of the retreatists either had mixed feelings about the decision or said that they disliked the idea, whereas most activists looked forward to it.[10] Retreatists may have felt that they no longer wanted to work or that they should retire, but they were less certain than the activists about how good it would be.

Some of their uncertainty may have been reinforced by their wives, who also were less likely than the women married to activists to have looked forward to their husband's retirement.[11] Nor did those wives who had mixed feelings or who were opposed to the idea hesitate to say so. They asked, Will he be bored? Will he just hang around watching every move I make? One went so far as to say, "Myself—I never went for retirement. I wanted him to keep working, but he was determined to retire."

Only 60 percent of the printers said that they had made any plans for retirement, and by plans they meant moving away from New York, taking a trip to another part of the country, or visiting relatives and friends. The activists who looked forward to retirement were the most likely to have made some plans, but whether the retreatists looked forward to it had nothing to do with their making any specific plans. This is further evidence that they entered retirement in a reactive rather than a proactive way.

Life-style in Retirement and Adjustment to Retirement

Even though as many as two out of five printers had made no plans for retirement, they were more prepared for leisure than one might have expected. Most had developed some interests and pastimes during their working years. Only one out of four said that he had become interested in something new after he had retired.

Activists were more likely than retreatists to have developed interests and hobbies, and the interests of the activists were greater in number and spanned a wider range. Their hobbies included breeding horses, playing the flute, cycling, and jogging, as well as more typical activities such as billiards, golf, travel, and reading. And, as this list suggests, the interests of the activists represented self-development as well as diversion.

Although both types of men regularly keep in touch with family members and equal numbers belong to a club or a fraternal lodge, activists are somewhat more likely than retreatists to spend time with friends and neighbors and to continue their leadership role in the union. Recall that, unlike retreatists, they were more likely to have been chapel officers or officers in their union when they were working. Now in retirement, they are twice as likely as retreatists to be sought out by fellow retirees for ideas, information, or help with a union matter.

It can be assumed that these men would not have taken the trouble to pursue a particular hobby or activity if it were not satisfying. However, the advice that retreatists and activists offer to the next generation of printers shows that they differ in the importance they attach to keeping busy in retirement. Almost one-half of the activists, but only one in ten of the retreatists, made a point of suggesting such things as "develop hobbies," "find things you like to do," "widen your circle of friends." Two of the retreatists were quite explicit that one does not have to be a "doer" in retirement: "People will tell you that you have to have a purpose in retirement. Remember, any purpose will do. Loafing is just as good as anything else." "If you get the idea that you always have to be doing something, you'll end up feeling guilty and miserable."

Although activists had a higher income than retreatists ($850 versus $750 median monthly income), this is not a critical factor in how they spend their time. Activists with an income below $850 report that they have as full a life in retirement as those who have a higher income. Almost four out of five of both activists and retreatists told us that they were living at least as well or better than just before they retired. Thus, differences in their use of leisure cannot be accounted for by differences in their retirement incomes.

Overall, the life-style of these retirees emphasizes that there is a marked continuity between preretirement and retirement life. Rather than create new roles or interests for themselves, they have stayed close to what they know and were like prior to retirement. In this regard they fit more closely the "continuity" theory of retirement than any of several others which have been set forth.[12] This theory presupposes that in adjusting to retirement, individuals seek to maximize the continuity between previous and present roles rather than to create new roles and engage in other meaningful activities (Rosow 1963; Atchley 1976). In other words, retirement does not change the orientation of men. Their aspirations, their interests, their engagement with life are much as they were before retirement.

Part of the price that retreatists paid for their life-style was that it took them longer to adjust to being retired. One-fourth said that it took them about a year, and one in five had still not adjusted two years later. Most activists adjusted within a few months.

Also there are differences between these two types of men in how satisfying they find retirement to be. Activists more than retreatists are largely pleased with their decision and have few reservations about it. But more than one-third of the retreatists wish they had not retired or are unsure about having made the right decision—a feeling shared by only 8 percent of the activists. Moreover, a large majority of the retreatists, but fewer than one-half of the activists, are glad that they did not retire sooner. One out of three retreatists often has times when he does not know what to do with himself.

The descriptions of retirement by these two types of men reveal how different the satisfied activists are from the uncertain retreatists. The latter want to talk about the past and say little about retirement other than that it is important to save money. Some are very blunt: "The best advice I can give a younger man is never retire," and they themselves wish that they could have retired gradually, rather than all at once. As one put it, "Maybe it would have been easier that way."

The perspective of the activist, who appears to be thriving in retirement, is quite different. The message he conveys to the younger printer is that although he enjoyed work, he takes its loss in stride and concentrates on the present. A recurrent theme throughout his remarks about the advantages of retirement is the importance of flexibility, resourcefulness, and the capacity to adapt to changing conditions.

> Life is what you make of it, not what you are afraid might happen if you quit work.[13]

> As for retirement—it's wonderful.

> I see how boredom gets some men fast, but not me, and it has nothing to do with age [a seventy-four-year-old].

> You have to expect to be disappointed in some things you had planned. Then you just make other plans.

Finally, wives must be considered in this profile of the activists and retreatists.[14] Earlier we pointed out that the wives of retreatists were more hesitant and uncertain about their husbands' retirement than were the wives of activists. It also took longer for the wives of retreatists to adjust to their husbands' retirement.

The different responses of the wives suggest that when a couple is well matched (that is, when their approach to the day and their pace of activities are similar), they tend to be happy and satisfied in retirement. When, however, opposites are paired, adjustment to retirement can be difficult. The data are only suggestive, but the insights into what retirement can mean to people who have been married forty or more years are instructive.

The first difference between these husbands and wives is the way they expressed themselves.[15] A few of the men made pointed statements about what makes for a successful retirement: "Don't move too close to your relatives." "Take only an occasional drink—scotch, if it is available."

Most men, however, spoke in generalities when they appraised their life in retirement and when they offered suggestions to the next generation of retirees about how to make it a satisfying experience. "Just be thankful you are able to retire." "Do not measure what you can do by age." "Plan to use your time sensibly."

In contrast, the women were more forthright and concrete about retirement, regardless of how happy they were. Those who were less happy described their lives and their adjustment to retirement this way:

> They don't adjust; you have to adjust to them. I have to be at his "beck and call." When he wants to go someplace, I have to go.

> I have things to do; he doesn't, and sometimes it drives me crazy.

> It is hard to be home so much. I miss working . . . the hustle and bustle. I find the day long. I wonder what I have accomplished.

> He's underfoot. I like having time to myself. I can't even visit my girlfriends.

> You have to have a lot of patience and understanding. You have to listen and then forget what you heard. You get a lot of criticism.

The women who were happy said:

> We both like taking life easy. A few errands, some TV, a trip to the adult center, visit the grandchildren, and we're happy.

> My husband loves so many things—music, sports, reading. I'm the same way. We have fun without spending money.

> We always did everything together. A lot of women complain, but I think that these are the women who don't like to stay home.

> I like having him around. He's easy to get along with. Everything is better from shopping for groceries to sitting in front of the TV together.

> We're together, and off we go!

The first set of comments identifies the strains a couple confronts in retirement when their interests diverge; and the second, how satisfactory retirement can be when the interests of both husbands and wives are similar. These differences suggest that just as there are two types of retired men, there are two types of retired women—the "homebodies" and the "extroverts." The former are typically women who have spent their lives raising children and taking care of their home. They like being at home whether they are attending to housekeeping activities, sewing, reading, or working the crossword puzzle. These women call themselves old-fashioned, and they report that they have enjoyed centering their lives on their home and being with their family. The extroverts, in contrast, find satisfaction in activities outside the home. They want to go places and do things; their life-style reflects independence more than dependence, and they see retirement as no less a time than any other for engagement with the world outside their home.

A few of these women had recently retired themselves, and for them being at home was monotonous. One took a part-time job after a year of retirement, because, as she said, "I felt resentful tied down. I like to socialize; he doesn't."

One might expect that satisfaction with retirement would be more difficult for the woman identified as a homebody, because for so many years she had the responsibility for decisions in the house and now in retirement she must share them with her husband. This may be so in isolated cases, but the account of retired printers and their wives suggests that compatibility and similarity in outlook toward the day are more the test of retirement than the ability to subdivide a role which once was fulfilled by one person.

Summary

This chapter has centered on the printers who retired, men who voluntarily decided to leave a steady, well-paying job because they preferred retirement. For this purpose we considered retirement not so much as a social event, but as a process which challenges an individual to undergo change as he adapts to a new and quite different set of circumstances than those he has left behind. The fact that these men chose retirement, that their income was adequate, and that they enjoyed good health (with the exception of those who retired because either they or their wives were ill) allowed us to examine the adaptation process free of some of the constraints which have characterized previous research on the attitudes toward and the experiences of retirement. Further, the information provided by both husbands and wives enabled us to consider retirement not just from the perspective of the retiree himself, but rather from that of a husband and wife.

Everything was more difficult for the involuntary retiree, the man who stopped working because of poor health, either his own or his wife's; and the opportunities for pleasure available to these people appear very limited, and their need for resources, both money and community services, including some which might be provided by their union, is considerable. These men and their wives face the problem not of learning how to stay healthy, but of how to live as enjoyable a life as possible despite their health problems.

An analysis of the experiences and attitudes of retreatists and activists reveal that they differ in their level of education, and that of their wives, as well as in the achievements of their children, their life-styles, and their aspirations for retirement. The activist is a man of higher achievement, more socially integrated, and happier in retirement than the retreatist. Many retreatists look back with yearning to the days when they were journeymen union printers; now in retirement, they experience a kind of social anomie. They see themselves as having slowed down, and they acknowledge difficulty being retired. Some are all too aware that they must be prodded to participate in their community, whether this means a card game or just going outside for a few hours. A few made a point of saying that in retirement you have to enlarge your group of friends, a task which they find difficult. If they are married to women who prefer to spend the day differently than they do, the adjustment process is further strained.

The men who find adjustment to be a particularly slow process are the ones who wish that they could have retired gradually rather than all at once. They appear to blame some of the problems they confront on having had to "cut the cord," as one put it, with such finality. It is not that they want to work some other place, but rather that they wish they could have continued to work part-time at the newspaper. Although their involvement with work was more passive than that of the activist man, now in retirement, they—not the activists—miss the camaraderie of the composing room. Perhaps retirement would have been easier for some if they had been allowed to taper their working days. We are thinking especially of those retreatists who have come to the conclusion two years after retirement that the decision was not a good one.

The experience of retirement has been more difficult and less rewarding for retreatists than for activists, but most retreatists do not regret their decision. In a very real sense, this is a tribute to an older group of men whose lives were anchored for so many years to work and to providing for their families. They had spent the majority of their years at a time in history when only the ill or the privileged stopped working. It was not until they were in their late fifties and some in their sixties that retirement became the norm. Their children, their neighbors in other lines of work, the dollar incentives of social security, and their pension rights awakened them to the possibility of retirement, and they decided to take advantage of it. Under the best of circumstances it is a considerable undertaking to make the transition from working to being retired, and for men of their background it is even more of an achievement.

Notes

1. There is a vast literature on the impact of these variables on the retirement decision. See, for example, Williams, Tibbitts, and Donahue 1963, (Volume 2); Bracey (1966); Simpson and McKinney (1966); Riley and Foner (1968, especially volume one); Barfield and Morgan (1970); Carp (1972); Jaffe (1972); Jacobson (1972); and Binstock and Shanas (1976).

2. Some may argue that although printers were willing to tell us why they decided to stop working and how they felt about work and retirement, these statements are of limited use in understanding what the retirement decision actually entailed. How can we know what they chose to remember or if, in fact, they can honestly detail how they came to the decision which they did? After all, the bonus period was 1974 to 1975, and we did not conduct the study until 1976.

Granted, the data are retrospective and subject to selective recall. By 1976 the attitudes printers expressed and the explanations they offered were probably often-repeated responses to the question Why did you retire? But they illustrate what they perceived to be salient factors. Moreover, there is a relationship between the explanations they offered, their attitudes, and their life-styles before they retired and their present involvement and satisfaction with life as retired persons.

3. Percentages add to more than 100 because retirees could give more than one answer.

4. Three did not answer the question.

5. Almost one-half of the wives whom we interviewed did not work outside the home once they married. The remainder are either working (18 percent) or are now retired (36 percent). Those who have stopped working gave as their major reason for their decision "My husband wanted me to retire."

In her book *Blue-Collar Marriage* (1967), Komarovsky indicates that patriarchal attitudes dominate in working-class families. Such attitudes may or may not extend to the retirement decision, but the statements by printers' wives suggest that they may.

6. Paul F. Lazarsfeld formulated this model, and his paper "The Art of Asking Why" first appeared in 1934 in a marketing journal which is now defunct. The article is reprinted in Lazarsfeld (1972). See also Ziesel (1968).

7. The typology was based on answers to three questions: the seven-item battery on reasons for retiring, the main reason(s) for retiring, and advice to prospective retirees, when these remarks included information about why printers had retired.

8. The opposing orientations toward retirement of the retreatist and of the activist parallel those of the traditional and instrumental printers discussed in chapter 3.

9. See Barfield and Morgan (1970). The difference between retired automotive workers and these retired printers in terms of subjective appraisal of health suggests that the latter faced more severe health problems and it was *the* reason that they retired when they did.

10. Virtually every retirement study includes one or more items on attitudes toward retirement. In our study, 56 percent of the retired printers looked forward to retirement, 32 percent had mixed feelings, and 11 percent disliked the idea. These data parallel those reported for automotive workers (Barfield and Morgan 1970) and for a sample of residents in the Greater Boston area (Sheldon, McEwan, and Ryser 1975).

11. Of the fifty wives of retired printers we interviewed, twenty-five were married to retreatists, twelve to activists, and the remainder to men who had retired for health reasons.

12. The literature on the relationship among aging, retirement, and adjustment is considerable. It includes a number of sociopsychological theories which seek to explain what makes for a good adjustment to retirement. Some theories emphasize the need to find substitutes for the satisfactions of work; others see retirement as a time of disengagement; still others focus on the importance of accommodation or adaptation; and a recent, more comprehensive theory suggests that attention should be directed to appraising one's personal goals and broadening and reordering them, if they had been narrowly centered on work for many years. For an explication and critique of these various theories see, for example, Friedmann and Havighurst (1954); Cumming and Henry (1961); Williams, Tibbitts, and Donahue (1963); Riley and Foner (1968); Neugarten and Havighurst (1969); Havighurst et al. (1969); Streib and Schneider (1972); Shanas (1972); Binstock and Shanas (1976); and Atchley (1976).

13. The work of Streib and Thompson (1957) and of Donahue, Orbach, and Pollak (1960) suggests that this attitude toward retirement makes for a more successful adjustment than an attitude which reflects fear.

14. In Simpson and McKinney (1966) see the section "Family and Retirement" (by A.C. Kerckhoff) for a discussion of husband-wife expectations and experiences in retirement. Both husbands and wives were interviewed in that study as in ours.

15. The men completed mail questionnaires, and the wives were interviewed for about twenty minutes by telephone. It is possible that these two modes of data collection contribute to the differential response style, that is, the men being more general and abstract and their wives being more vivid and concrete in their comments about life in retirement. Previous research, however, provides evidence that more liberal and freer responses are obtained by mail than either by telephone or in person (Hochstim 1967; Wiseman 1972).

Our impression of the data obtained from retired printers is that they were direct in their response to specific items (say, activities which interest them and attitudes toward work), but less so in answer to questions such as assessing retirement and offering suggestions about retirement to printers still working.

Appendix 5A

Table 5A-1
Selected Differences among Retreatists, Activists, and Involuntary Retirees

	Retreatists ($N=36$)	*Activists* ($N=27$)	*Involuntary Retirees* ($N=21$)
Average age at retirement	69 years old	68 years old	69 years old
In good or very good health	88%	74%	40%
Considering income and expenses, living as well as or better than before	81	74	52
Monthly income is enough or more than enough to live comfortably	86	96	65
High school graduate or some college	14	32	35
Sent one or more children to college	49	79	65
Officer or candidate for office in Local 6	8	26	24
Favorable attitude toward retirement	58	74	29
Advise future retirees to have hobbies and activities	11	44	38
Glad did not retire earlier	70	46	84
Often have times when do not know what to do to keep busy	31	11	42
Took a year or longer to adjust to retirement	47	22	52
Often wish had not retired	37	8	30

6 The End of the Printers' Union

The two most important forces shaping the working lives of printers at the *Daily News*, *New York Post*, and *The New York Times* are automation and their union. By virtue of its strength and resources, Local 6 has skillfully engineered the transition to an automated composing room. Jobs have been guaranteed; the wage scale allows for cost-of-living and productivity increases; and voluntary, not mandatory, retirement obtains. At the same time automation has triggered a chain of events which will, in all probability, end the life of Local 6 as a single-craft union. In this chapter we consider the critical economic and organizational issues which this union confronts and the response of its membership to the change which is taking place in the union.

Automation and the Union

Automation poses direct and indirect threats to the viability of trade unionism in the printing industry. First, automated technology has greatly reduced the number of jobs in printing. Between 1969 and 1976 alone, the number of employed union printers in the United States and Canada declined by 24 percent. And, although Big Six continues to be the largest of all the International Typographical Union (ITU) locals, its membership is decreasing rapidly. In 1962 there were 9,559 active journeymen, but by 1977 there were only 5,233 (*Local 6 Bulletin*, October 1977). Moreover, the sharpest decline has been in the newspaper branch of the industry. Between 1962 and 1977, the number of printers employed there has declined 55 percent, from 3,771 to 1,673. Some of the explanation is attributable to automation and some to the closing of several daily newspapers in the city.[1] By the time the contract expires in 1985, it is estimated that only about one-fourth of the printers will still be working in the three composing rooms, and no new printers will take the place of those who have left. The greatly reduced workforce will, the experts agree, be more than sufficient to put out the papers.

Second, printers and their union officers recognize that the sweeping technological changes in the composing room have reduced their bargaining strength. The degree to which this is the case became dramatically clear at

This chapter was written in collaboration with Jill Bernstein, a member of the study staff.

the *Daily News* early in 1974. With the composing room workforce on the picket line, thirty-five employees (largely secretaries) were able to turn out over 2 million copies of the paper each day. The problem was aggravated by the fact that the ITU did not support the walkout and that some of the other unions at the paper did not respect the picket line.

Third, with no new printers being employed at the newspapers, the proportion of older workers and pensioners on the membership rolls has increased sharply. Over one-fourth of the total membership in the ITU were pensioners in 1976, and the figure is rising. Moreover, the age structure of Local 6 is particularly skewed in favor of older printers. In 1976 the average age of all journeymen in the local was 48.7, compared with 46.4 for all ITU members (*Local 6 Bulletin*, May 1977). At the New York newspapers, where there is no mandatory retirement, as there now is in the book and job (commercial) shops, the working printers are even older. According to 1976 records of Local 6, the average age of the men at the three papers was fifty-six.

The gravity of the problem of an aging membership is clear to the rank and file as well as to union leaders. One printer put the issue quite succinctly:

> We have a real problem here. Now that there are no new members coming in anymore, and older people are retiring in greater and greater numbers, the pension funds are being thrown out of kilter. We [younger] workers are supporting growing numbers of pensioners. Sure, you can talk all you want about fraternalism and "I am my brother's keeper," but the changing balance between working and retired members is something that has to be looked at without emotion.

Automation, together with the concomitant reduction of the workforce and the increasing obsolescence of printing skills, not only has exacerbated the organizational and fiscal problems of the union, but also has resulted in a number of specific constraints at the job site. The new technology undermines the "sacred pillar of priority," the seniority system. Printers are assigned priority according to the date on which they deposit their card at the newspaper, and historically their priority rating has been the criterion for job assignment, shift selection, slide days (days off), and vacations. As one chapel officer told us, "Priority is everything."[2]

In the past, for example, if an operator was needed for a new assignment, the one with the highest priority would have first choice. Today, however, the new technology has eliminated many of the functions which differentiated assignments in the composing room. Proofreading can be done on VDT equipment, and this eliminates old distinctions between operators and proofreaders. Either handmen or operators can be retrained for many of the keyboard and nonkeyboard operations.

Before the new contract, priority was the key factor in selecting printers for retraining. Even after the ratification of the 1974 contract, which gave management discretion to select printers for retraining regardless of priority, at least at one of the large dailies there was a tacit agreement that if management selected a low-priority man for retraining, it also had to give the opportunity to one with high priority. The union recognizes that unless a new shopwide system of priority is instituted, incorporating a workforce trained in all aspects of the new automated processes, the traditional authority of the craft in protecting individual printers from the arbitrary discretion of management will be permanently ended.

An even more serious issue resulting from the reorganization of production in printing is that printers are increasingly involved in direct competition with other newspaper personnel for job assignments.

> The jist of the problem . . . is that people not privileged to do the work are doing it—that is, people who are not members of our union are doing it. For example, editors are making typographical corrections. We want to hold proofreading jurisdiction. Though our jobs are guaranteed, we don't want to relinquish any of our work. The guiding line is, whatever they [nonunion people] read and prepare, *we must read.*

A jurisdictional dispute between Local 6 and the Newspaper Guild arose over the assignment of the classifieds processed on the optical character-recognition devices (OCRs). The National Labor Relations Board was called in, and eventually Local 6 was granted jurisdiction over OCR work. But disputes of this nature arise regularly as more and more processes are automated. Similar jurisdictional disputes have arisen between compositors, who have traditionally set up the pages of the newspaper, and the paper's art department. With the new equipment it is less clear than ever as to what is "type" and what is "art."

Job jurisdiction issues are closely related to a third organizational problem faced by the union. In order to survive, the ITU must reverse the precipitous decline in its membership. "We live with automation," the printers have written in their *Typographical Journal*, "but we grow by *organization*." Further,

> Organizational efforts should include all unorganized employees performing work pertaining to the graphic arts and communications industry—both non-mechanical and mechanical departments—in other words, from front door to back door, where no other unions are involved.[3]

An organizing committee held its first meeting in April 1977 in New York City. With the help of the New York City Central Labor Council, the committee announced plans to embark on a large-scale campaign. "The union," the committee members declared, "is prepared to take direct economic action by paying $100 a week benefit to workers who need help in organizational activity" (*Local 6 Bulletin*, April 1977).

The problems of organizing new shops are many in New York City. Special provisions need to be made for new members of the union who would not be covered by the guaranteed income or job agreements that now prevail in the local. Further, the existing benefit package may not be all that appealing to prospective members. The present membership is itself somewhat critical of the fringe benefits offered by their union, particularly in light of the high dues structure (4.5 percent by the fall of 1977). Several remarked that their wives who were employed enjoyed better benefits.

Another impediment to organizing new shops is the fact that Big Six printers have the highest wage scale of any local in the ITU. In 1975, for example, this scale was 31 percent higher than that of printers directly across the Hudson River who were members of the Newark local. Accordingly, employers can be expected to raise considerable opposition to attempts to unionize their workers under the aegis of Local 6.

A further constraint that the local faces in its attempts to organize nonunion shops in New York City is that, while the number of small-scale printing shops has increased, the services that they provide do not require the traditional skills of journeymen printers. Accordingly, these shops are unlikely candidates for union organizing efforts. Finally, there is an increasing trend for employers to either move their existing facilities or establish new ones outside of the Local 6 jurisdiction. The short-lived *New York Trib*, for example, was produced from the outset in New Jersey.

A more promising solution to the union's problems is to merge with one or more other unions. Leaders of the ITU have declared (*Typographical Journal*, April 1976):

> The *only* realistic answer for employees who work in the industry, and unions which represent them, is the *physical merger* of as many of the graphic arts and communications unions as possible. No other single action by individual local unions, or by an International, can provide the immediate salutary impact upon the strength and welfare of union members as can merger.

At the 1977 convention, the ITU's Committee of Graphic Arts Merger recommended that a blueprint be drawn up for merger with the Newspaper Guild (and the Guild also approved such an effort). Merger talks have also been held with the Mailers Union, the Paperworkers, and other printing-related unions.

Interest in the prospect of merger exists at the local level, too, and some Big Six unionists predict that merger with the Newspaper Guild might be effected as early as autumn of 1978.[4] Because Local 6 is the largest of the ITU locals, successful efforts to merge in New York City might well pave the way for merger at the international level.

While printers do not look forward to the end of their old, respected single-craft union, which they worked so hard to build, they reluctantly agree that merger is essential. They recognize that it will mean new priority systems and that benefit programs will have to reflect increasingly complicated differentials in the earnings of members engaged in a wide variety of jobs. And they wonder whether merger will necessitate a more complex, centralized bureaucratic structure which will widen the distance between the rank and file and their union leaders.

Printers View Their Union

While many of the men are aware of the problems facing their union, both working and retired printers remain loyal supporters of the ITU. The men at the *Daily News* and *The New York Times* overwhelmingly ratified the 1974 contract by a vote of 1,009 to 41, and six weeks later the same contract was ratified at the *New York Post*, 165 to 10.

Many printers express admiration for the ability of Bertram Powers, president of Local 6, to negotiate the present contract: "The new contract is a beauty. The question is: How did he do it without a gun?" Others, however, recognize that the economic rewards of the contract—lifetime job security and guaranteed cost-of-living adjustments—represent a tradeoff for management's full discretion to automate: "We had to give up a lot for what we got. Now they can automate the whole composing room and switch us around without asking us. But at least we got job security."

Printers point with pride to the democratic character of their union and for a variety of reasons:

> The ITU is a very democratic union. Everyone is equal. I mean everyone makes the same amount of money whether he is a proofreader, a makeup man, or whatever. Even the substitutes make the same as all of us. Like vacations—if a printer is here one year, he gets four weeks vacation; and I'm here twenty-nine years and I get four weeks.

> Our union is very democratic—they're always asking us to vote on some referendum or another.

Some printers also called attention to the fact that the books of the union are open for all to see. The monthly *Local 6 Bulletin* includes an accounting of all financial transactions conducted during the previous month, such as salaries and expenses and petty cash disbursements.

The high regard in which the printers hold their union is reflected in the fact that almost one-half of the working printers and two out of five of the retirees think that, compared with most other unions, the ITU is "the best." Fewer than one in five of both groups rate the ITU "about the same" or "not as good" as most unions. Further, there is fairly high consensus among age groups on this question: Younger printers are only slightly less likely than older men to rank their union at the top.

But printers do want improvements in the benefit structure, particularly medical benefits to cover those who retire before they are Medicare eligible.[5] When we asked the printers, for example, if "there were any things Big Six should be doing that it's not doing, or that it should be doing more of," more than one-half cited the need to improve the current benefits. They also emphasized the need of the local to intensify its efforts to organize more nonunion shops and to merge with other unions such as the Mailers or the Newspaper Guild: "They've got to organize more shops, bring more people into the union, more new blood." Another said, "There's talk of merger with the Guild. I'd like to see that because it's the only thing that will save us." Many said that although they would like to see Big Six doing more for its members, "It's hard—you can't ask for the moon and expect to get it. There are only three newspapers left in the city; you have to be thankful you have a job."

Goldthorpe et al. (1968, p. 114) suggest that the orientation of workers toward trade unionism reflects their general orientation toward work. Indeed, underlying the comments of printers as they talk about their union are two different perspectives, reminiscent of those which differentiate the responses of the printers to the new technology.[6] Traditionalists among the printers are more likely than instrumentalists to view their union as more than an economic benefactor. For example, almost one in five traditionalists is a member of social clubs or sports groups sponsored by the union, while this is true for fewer than one in ten instrumentalists. Moreover, it is not that traditionalists are "joiners" and instrumentalists are not, because instrumentalists are more likely than traditionalists to belong to clubs or groups which are *not* part of the union.

The comments of traditionalists and instrumentalists themselves reveal how differently each thinks about his union. Said one traditionalist: "If there is one thing the union should be doing more of, it's involving the men more in social activities. In the old days there was a lot more of that, and it pulled us together." He then added: "The union's taken good care of us all these years. Some of the guys don't appreciate it. Sure there's always room for improvement, but I think they're doing a hell of a job." Another commented: "The ITU used to be the best in the world. I wish we could make it like it was way back when everyone was together."

The instrumentalist, in contrast, has a pragmatic view of Local 6:

> The union should get more involved in the cold-type process; it should get together more with the publisher to work on the training process.

> For the amount of dues we're paying, our benefits could be better. If we want to organize more shops and increase our membership, we've got to look more attractive to the unorganized printing shops in the city.

The approach of the instrumentalist is perhaps best summed up by the printer who said, "When it comes down to it, although it's basically a union, I look at it [the ITU] as a lifetime employment agency."

Cleavage on the Composing Room Floor

At the time of our study, Big Six was caught up in several issues which result from the decline in the number of working members and the increase in those who are, or soon will be, pensioned. The problems of pension funding and the role of pensioners in the operation of the union were, as several said, "the hottest issues to face the union in thirty years."

The major program providing union printers with retirement pensions is the Negotiated Pension Plan, funded by employer contributions and union dues. In 1969, when Local 6 merged its own negotiated pension with that of the ITU, the Local 6 plan had liabilities which exceeded its reserves by about $4 million. As the deficit grows, Big Six becomes an increasing burden on the ITU pension fund.

The Fraternal Pensions

A second source of retirement money was provided, until 1975, by what the printers refer to as the ITU Fraternal Pension, funded by a 2 percent tax on members' dues.[7] Again because of the growing number of pensioners, the reserve funds of the fraternal pension began to shrink, and the monthly payments to retirees were reduced from $100 to less than $50. By 1975 it was clear that either the reserve funds would have to be increased by raising membership contributions or the fraternal pension would end. At the annual ITU convention in the summer of 1975, a motion (known as Proposition 128) was passed to reduce the pension fund floor by $1 million, gradually reducing the pension payments and eliminating them altogether when they fell below the $10/month level, and to change the ITU membership rules so that members could opt out of contributing to the fraternal fund altogether. "Associate members" would no longer pay 2 percent of their dues toward the fraternal pension and would forfeit entitlement to such a pension and to voting rights in matters relating to it. The leadership of Big Six was united in its opposition to Proposition 128. A statement by Big Six Secretary-Treasurer Kopeck was typical of the Local 6 officers' stand (*Local 6 Bulletin*, September 1975, p. 5):

> . . . Now, the fraternal aspect in our union is very, very important to the survival of the union. By extracting the fraternal aspect you are changing the very essence of the union and the very reason that this union has existed for as long as it has existed, through all the vicissitudes of our society and

> all the problems that we have overcome. Most of the reasons it has gone through this can be traced directly to the fact that we have always taken care of our own. We have done it through depression time, we've done it when there wasn't enough work to go around. We have always shared with those who went before us and gave us a union so that we could enjoy the fruits of our labor. This is very elementary and I don't think anybody would disagree with it if you put it just in those terms. But now we are being told that in order to save this union we should eliminate that, and if you eliminate that you have a husk of a union. You have a name but you do not have the union anymore. . . .

The referendum, however, was passed by the ITU by a majority of 5,000 votes on November 19, 1975, although it was soundly defeated (10 to 1) by the members of Local 6.[8] Less than three months later, on February 4, 1976, the members of Big Six passed a referendum by a majority of more than 2 to 1 (4,719 to 2,287) instituting a local benefit in place of the almost defunct ITU fraternal pension.[9] This resolution provided that when the ITU fraternal pension and the 2 percent dues to support it were discontinued, members of Big Six would begin to pay 2 percent dues to the General Welfare Fund of Local 6. From this fund, $50 per month would be paid to all Big Six members who had been on the rolls of the ITU fraternal pension, as well as to those printers who retired by June 30, 1977. Gradually, as the number of eligible printers decreased, this additional assessment would be lowered by 0.5 percent until it was phased out completely (*Local 6 Bulletin*, February 1976).

The officers of Big Six saw the endorsement of the new local pension as having two major implications. As then Secretary-Treasurer Engler said (*Local 6 Bulletin*, December 1976, p. 8),

> The fact that fraternalism is alive and well in Typographical Union No. 6 has been historically established and that it is an ongoing characteristic is proven by the initial mailing of the $50 Special Benefit checks to all eligible pensioner members of the Local who were on the ITU Old Age Pension roster at the time it was phased out.

Fraternalism was also the keynote sounded by President Powers, as he urged the membership to support the Local 6 special benefit (*Local 6 Bulletin*, January 1976, p. 1):

> I recommend a vote in favor of the local benefit proposition to be decided by referendum on February 4 . . . the average cost to the active member to provide this local benefit for pensioners would be one percent . . . a small price to pay to give a great boost to those men and women we worked with over the years whom we now call pensioners.

A second, if unintentional, implication of the Local 6 fraternal pension was noted, namely, that it would serve as an inducement for the older printer to retire (*Local 6 Bulletin*, April 1977, p. 7):

> It is expected that the cut-off date for the Local Special Benefit will act as an incentive for retirement. A member who is 65 or older would have to work an additional five years to generate another $50 (a month) in pension benefits from the ITU Negotiated Pension Plan.

Some of the printers, however, expressed anger at the fact that the union itself, through the 2 percent tax on its members, was financing management's needs for attrition of the workforce.

An analysis of the voting patterns of our sample on the issues of Proposition 128 and of establishing a Local 6 fraternal pension reveals the cleavage in Big Six between those who maintain a traditional, fraternal stance and those who have adapted a more pragmatic, instrumental position. The dividing line is between those who have reached an age at which they, too, may be contemplating becoming a pensioner (sixty or over) and those who have not. Just one-half of the printers under forty-five, but 85 percent of those over sixty, voted in favor of continuing the ITU fraternal pension. Similarly, the older men were more than twice as likely as the younger printers to approve the additional tax on their wages to establish a Local 6 fraternal pension.

Percentage Voting to:	*Under 45*	*45-59*	*60 or Older*
Continue the ITU fraternal pension	50	56	85
Institute a Local fraternal pension	41	45	86
N	(132)	(122)	(150)

That one-half of the youngest (under forty-five) printers are ready to dig into their pockets to take care of the pensioners indicates that when it is a matter of "I am my brother's keeper," traditionalism still exerts strong fraternal pressures—even among the younger men. And indeed, the printers' orientation toward work, traditional or instrumental, sharply reflects their positions on the two issues. Three out of four traditionalists, but only one-half of the instrumentalists, voted to continue the ITU fraternal pension or to institute the local one. Moreover, examining the vote on the Local 6 fraternal pension, we find that at each age level traditionally oriented printers were more likely than instrumentalists to endorse a 2 percent tax on their wages for the purpose of supporting the retirees of Big

Six.[10] Even among the youngest printers, three out of five traditionalists, but just one in four instrumentalists, took a fraternal stance.

	Percentage Voting Yes on Fraternal Pension	
Age	*Traditionalists*	*Instrumentalists*
Under 45	61 (41)	27 (45)
45-59	71 (35)	35 (43)
60 or older	95 (39)	78 (56)

Support of the local pension rested largely on a fraternal basis, as the printers' comments indicate:

> Men who supported the union through thick and thin should be supported now. The union has to take care of its own; that's the true sense of unionism, and I believe in it.

> Who established this union if not the pensioners? I think they deserve a fair shake, and they've gotten shortchanged. The pensioners are the ones who made it a bed of roses for those young fellows, and now the young fellows should pay for it. It's not even that much money.

> All those older men put in forty or fifty years, and they're entitled to the pension. If we don't take care of our own, who's going to do it?

The instrumentalists who supported the local benefit were less nostalgic: "I'm single, so I can afford to be generous." "It just takes a little bit of money out of my pocket, so why cut them down?" By and large, however, the instrumentalists—particularly those under sixty—voted against the local fraternal pension. Many were concerned about the shrinking working membership and the rapidly growing cadre of pensioners in Big Six.

> The only way we can survive is through amalgamation with the other unions or by organizing nonunion shops. But the fraternal pension killed that opportunity. Look at it this way. When these guys ask what are the benefits of joining up with Big Six, we can probably convince them that it's to their advantage. But when they ask what they will have to *give*, it's another story. When they find out they'll have to kick in that 2 percent to take care of the old timers, and that they will never get anything back from this money, they'll simply say "forget it!" This is how we younger men feel, and that's why we oppose the local fraternal pension.

Other instrumentalists took an even more pragmatic stance:

> Why should I pay into something that I won't get any return on? I have a family to support—I don't have to pay for the pensioners' condominiums in Florida.

> No—I voted against the 2 percent pension. It's an extra $7 or $8 from my take-home pay each week, and I'm not even covered by it. Maybe it sounds selfish, but I have to think of myself.

Three out of four of the older instrumentalists voted in favor of the Local 6 pension benefits. The rationale for their support was simply "I'll be retiring before the cutoff date [June 1977], so I'll be getting something out of it for myself."

Issue of the Pensioners' Vote

A second divisive issue centered on the voting rights of pensioners. Repeatedly, younger printers expressed anger that pensioners turn out in full force to vote on issues which affect the lives (and wallets) of working printers. Over the years the ratio of working to retired printers has changed drastically. In 1962, it was 5.2 working to one retired printer, and by July 1977 there were only 1.7 working printers for each retired printer (*Local 6 Bulletin*, October 1977, p. 22). Apart from the financial burden that this changing balance between working and retired printers has placed on the workforce (as well as the fear that the negotiated pension reserves will be largely dissipated by the time the younger printers retire), the younger men have become increasingly angry about the voting power exercised by the increasing number of retirees. As one noted, "They stack the meetings; it's gotten to the point where they can carry any election."

In our sample of working printers the differences were clear-cut and dramatic about whether all members of Big Six should vote on every union issue. One-half staunchly upheld the right of the pensioner to vote on all union matters:

> It's their union. The fact that they've retired from work doesn't mean that they're retired from living. Union matters should be weighted by minds that are older and experienced. They've been in it thirty, forty, and fifty years. It's their life.

> Without these old-timers, where would we be? It took the old-timers to get whatever we now have; they taught us all we know. They've paid dues all these years—they're entitled to a voice.

The other half saw the issue very differently:

> It's my money they're voting on, and I don't like that. It's the guy who's working who should have the full voice.

> They're out of the trade now. Their problems aren't my problems. Besides they vote as one bloc, and they're getting bigger and bigger. They shouldn't be able to vote on things they don't know about and that affect *us*.

As in the votes on the Local 6 and ITU fraternal pensions, the cleavage was largely between the older and younger men, with three out of five of the older men, but just over one-third of the younger ones, supporting the pensioners' vote on all matters. Within each age group, however, again the printers' orientation differentiated men on this issue.

	Percentage Voting Yes on Pensioners' Right to Vote	
Age	*Traditionalists*	*Instrumentalists*
Under 45	43 (47)	24 (51)
45-59	50 (38)	30 (47)
60 or older	67 (45)	56 (57)

Thus, the strong fraternal spirit, long characteristic of this union, continues, even among younger printers, and serves to maintain the commitment to fraternal programs despite the severe threat to the union's financial reserves.

The issue of the pensioners' vote was eventually resolved with the help of the pensioners themselves. To quote President Powers:

> . . . the pensioner members themselves . . . do not wish to let this issue generate friction between the two groups of members [journeymen and retired printers] because at their last meeting they voted unanimously to support any resolution by Local 6's ITU delegates which would prevent pensioner members from voting on scale negotiations, contracts or any question affecting active members only.

Powers applauded the pensioners' move and called on all members to support such a resolution "in order not to put too much strain on the relationship between generations" (*Local 6 Bulletin*, July 1977, p. 3). On November 16, 1977, this resolution was put to a referendum vote, and it was passed resoundingly (5,473 to 594) by the total membership, marking yet another accommodation by the union to the changing conditions in the industry which threaten its survival.

Big Six and the Future

Despite the fact that some of the old-line fraternalism of a craft union continues in Big Six, there are indications that Local 6 is taking on the character of an industrial union. With the security, for example, of an eleven-year contract in the newspaper branch of the industry, and having resolved the most pressing conflicts between pensioners and journeymen, Local 6 is turning to the larger issues of reorganization and merger. Within the next few years it is likely that merger with one or more of the printing-related unions will be effected.

As a result, the printers' union of the future will be a larger, more centralized bureaucratic organization, geared for coordinating the activities and benefits of a stratified workforce. There are signs that Big Six is already beginning to take on organizational characteristics typical of many industrial unions. It has a well-entrenched political party which has controlled Big Six politics for more than a generation, there is evidence of increasing membership apathy, and it has begun to concern itself with national and state politics.

In 1976 through its Political Action Committee (not formed until 1974) Big Six broke with tradition and officially endorsed Carter for President, as well as the New York Democratic candidate for the U.S. Senate. The committee also voted to contribute to the political campaigns of the candidates that Local 6 endorsed and urged its membership to do likewise. Members were told that the local's decision to endorse the national Democratic ticket "serves to underscore that its members feel that political action is necessary to solve some of the problems facing them" (*Local 6 Bulletin*, October 1976, p. 1).[11] These political activities have been coordinated through the Committee On Political Education (COPE) of the New York State AFL-CIO and signals that Big Six is joining with other industrial unions in seeking solutions to its problems through political influence.

The ITU has long been the exception to a second general characteristic of large-scale unionism, namely membership apathy. Union activities are still legitimated by referenda of the entire membership, and voting on both the local and international levels remains high.[12] Still, there are signs of waning membership participation in union affairs. For example, in the study of Big Six printers almost twenty-five years ago (Lipset, Trow, and Coleman 1956), it was reported that 80 percent had attended at least one of the monthly meetings of the local during the previous year. In contrast, only 59 percent of the printers in our sample told us that they had attended at least one meeting. As one printer said,

> Everybody votes, but no one comes to meetings.[13] Even the major meetings—the ratification meetings—don't get large turnouts. There were only about 800 at the last regular meeting out of a membership of 7,000. The book and job printers didn't even show up when their own contract was up for ratification.

The decrease in membership participation may reflect a general feeling on the part of the printers that they have little influence on how things are run in their union. A total of 40 percent of the printers in the Lipset, Trow, and Coleman study, but only 8 percent of those whom we interviewed, felt that people like themselves had a lot of influence on how the union runs things.

Finally, one of the outstanding marks of the ITU throughout its long history has been its two-party system, unique in the annals of union organization. At least at the local level (in New York City), however, the two-party system is a thing of the past. But for the position of secretary-treasurer, traditionally held by a member of the Unitypo faction, the leadership of Big Six has been in the hands of the Progressives for several decades. The opposition at the local level is the Unitypo faction, which is seen by many printers as a watchdog group rather than as a meaningful rival party.[14]

Twenty-five years ago, one-third of the Big Six printers reported that they favored the local's Progressive Party, and an equal proportion favored the Independent Party; the remainder said they were nonpartisan.[15] Today three out of five printers call themselves Progressives and only one in twenty endorses the opposition party, with the remainder claiming that they "vote for the man or the issue."

By "the man" printers generally mean Bertram Powers, who has been the president of Local 6 since 1961. A minority of dissident printers complain of "one-man rule" and liken Powers to "a Caesar." But overall, his leadership is popular with the rank and file because he has "delivered" in the face of serious obstacles. Much of the credit for the 1974 newspaper contract is given to Powers' considerable bargaining skills. He also enjoys the enthusiastic support of pensioners.

The critical issues facing Big Six today, however, are not those of wages and job security, but rather those of survival. At question is whether Powers and the progressive leadership will be able to effect a merger even though it will mean transforming the union into a different type of labor organization. The loss of the status of a prestigious single-craft union and the traditions and customs it has developed will obviously be painful for the membership, and their leaders are aware of the overwhelming changes they must deal with and the resistance they may encounter. As they told the members (*Local 6 Bulletin*, August 1977, pp. 1, 3),

> If merger is to be finally accepted and consummated, our members must be prepared now to accept many changes. What must be kept uppermost in our mind is that the substance of the merger is what is important—not the format.
>
> Our cherished name will almost assuredly have to be changed. Certain international offices may or may not accrue to members of our union. Many traditional laws, customs, and institutions will undoubtedly have to give way to new practices and procedures in order to reflect the new, larger and more comprehensive union.
>
> As difficult as it may be to accept such changes; as strong as sentimentality, custom and tradition is imbedded in our membership, we believe our members must be, and are, prepared to face such changes in order to construct and build a union that can better protect and improve our living standards, bring the benefits of unionization to other workers in our industry through organization, and meet the onslaught of the employers on our jobs and conditions.

In conclusion, changes in the printing industry will mean changes in the organization and operation of this union. If Big Six can succeed in perpetuating the tradition of democratic unionism in the complex and bureaucratic world of industrial unionism, it will be not only a tribute to its leaders and members but also an important lesson for students of industrial relations as well.

The struggle of the ITU and, more particularly, of Local 6 is not yet over. Its succeeding efforts will say a great deal about the dynamics of democratic organization in the face of the structural constraints of economic and technological change.

Notes

1. Only three major newspapers remain in New York City today, compared with six in 1954.

2. With the closing of the books, a printer's priority cannot change. Those at the bottom of the priority list are "locked in" to the less desirable shifts, slide days, and vacation periods. According to one informant, "There's some talk about breaking the vacation into two periods, so that the first two weeks would be assigned starting from the top of the priority list and the second two weeks starting from the bottom. Frankly, I don't think it's fair for them [the low-priority men] to make a demand like that—they're lucky to have a job."

3. *Typographical Journal*, April 1976, p. 5. Between 1973 and 1976, eighty-five ITU locals in the United States and Canada won bargaining rights through local labor boards, while forty-five locals failed to obtain recognition. But this success rate "is not enough to offset the number of members lost through attrition and death, those leaving the printing industry and/or the absence of apprentices coming into the printing industry through our apprenticeship program" (*Typographical Journal*, April 1976, p. 6).

4. As of the autumn of 1979, such a merger had still not taken place.

5. Pre-Medicare coverage for the retiree has since been added to the benefits package.

6. See chapter 3.

7. Technically, this is not a pension but a "special benefit fund."

8. The vote against Proposition 128 at New York Local Number 6 was 6,704 to 698 (*Local 6 Bulletin*, December 1975). A vote for the proposition signaled the readiness to see the ITU fraternal pension phased out; a vote against the proposition indicated disagreement with its being phased out.

9. In the local as a whole, the printers in the book and job shops voted to drop the fraternal pension while the newspaper branch voted to continue it. The outcome was essentially determined by the overwhelming vote of the retired printers to continue the pension.

10. The same pattern held when we examined the joint effects of age and orientation toward work on the printers' vote on Proposition 128.

11. Also 1976 was the first year that the ITU endorsed a national slate.

12. Even this limited role of the rank and file may be eroding. For example, in 1976 the membership was asked to approve a new policy whereby the ITU Executive Council would be empowered to change the rates for officers' car travel allowances without the approval of the entire membership.

13. Voting takes place in the shop.

14. The Unitypos originated as a faction advocating the ownership and operation of printing establishments by the printers' union itself.

15. Data are from the files of the Lipset, Trow, and Coleman study of the ITU.

7 Epilogue: Ghost Town

It has been three years since we began our study of printers in the composing rooms at the *Daily News, New York Post*, and *The New York Times*. At that time the eleven-year contract between Local 6 and the Publishers' Association was only two years old, and the goal of automating the three major dailies was yet to be achieved. The plans were on the desks of the publishers and on the blackboard in the foreman's office, but only some of the equipment was on the floor. The changeover was proceeding faster at the *Daily News* and *The New York Times* than at the *New York Post*, but estimates varied as to how long it would take to fully automate the three composing rooms.

The *New York Post* was the first to accomplish the task and in only three months' time. By June 1978 it had converted completely to automated typesetting and to the high-speed production of printing plates. Over the July 4 weekend of the same year, *The New York Times* did likewise. In June 1979 the *Daily News* shut down the last of its linotypes, but it will be another year before the full system is operational.

The high tables at which aproned handmen once stood, deftly manipulating metal slugs to compose a page, have been removed. Some of the old, lumbering linotypes have been sold; the remainder are crowded together in a darkened corner of the composing room aptly known as "Ghost Town" at the *New York Post*. At this paper one linotype has been made into a museum piece. Cleaned and polished, it stands in a prominent location on the fourth floor, a constant reminder of days gone by. Similar plans are underway at the *Daily News*.

The noise, the smell, and the grease are gone, but so also are a lot of other things, including work for printers. Proofreading is virtually obsolete, except for scanning a few advertisements and maybe an editorial. Sufficient checks are built into the computerized billing system so that almost no proofing is needed. Paste-up of advertisements and the printed page comprise most of the workload, but this is also temporary. By the spring of 1980 *The New York Times* expects to install new equipment that will computerize the page layout process, thus eliminating paste-up. Only the machinists—"from grease monkey to white-collar man," as one described himself—foresee no shortage of work. Now called technicians, machinists service all the VDTs in *The New York Times* building. For most printers, however, the problem is not enough work, and there will be even less in the future.

In the beginning, automation was exciting and challenging. There were new developments practically every day, and many signed up for retraining to prove to themselves and to everyone else that they could learn the new skills and be part of the new order. Now most of the retraining has been completed, and much of the work is routine. "No noise, no hustle and bustle. You just sit around and watch it [the finished copy] come out of a box." Some older men are on the sidelines waiting to retire; only a few plan to come to work indefinitely. But retrained printers face problems, even though they have adjusted well to their new work. Their options are much the same from paper to paper, but the present tempo of work and their mood differ somewhat depending on whether they are in the composing room of the *Daily News, New York Post,* or *The New York Times.*

Printers at the *New York Post*, the oldest continuously published daily in the country, are the least disgruntled. For years they feared that their paper might shut down at any time, because circulation was poor, advertising revenues insufficient, and the publisher near retirement herself. But a new publisher took over in January 1977, and the *New York Post* is reborn. Circulation has increased, and advertising revenues are expected to do the same. Instead of being stacked only at newstands and in candy stores, the paper is now out in the open. Machines dispense it at street corners, and teenagers sell it at toll booths, on Madison Avenue, and even on Victory Boulevard in Staten Island where it had never been sold before.

The new publisher has unlimited energy and works twenty hours a day, the printers say. He goes down to the composing room to talk to them. If they offer a suggestion, he listens; if they have a request, such as for a new coat of paint, the job is done two days later.

Most printers at the *New York Post* have been retrained for all phases of the cold-type operation. That there is still work for the men in the composing room is partly because about ten on each shift are now doing photoengraving, work which had previously been farmed out. This work exists, because unlike at the *Daily News* and *The New York Times*, there is no Photoengravers' Union at the *New York Post*.

At the *Daily News*, "it's a sad scene," said one informant, pointing to the corner where forty dusty, old linotypes stand silent. The paper has yet to be fully automated, so there is enough work to go around. It is produced completely by cold type, but a considerable amount of the keyboarding of classifieds and editorial material is still being done in the composing room. Of the workforce 80 percent have been retrained, and for the most part printers have adjusted. "If they don't, they'll be left by the wayside." Thus far, there are enough stations for the retrained, but largely because the switch to cold type took place in early summer, a heavy vacation time. "When the men come back, it will be a different story. We'll have to start sending them out on productivity leave to keep the workforce in line with the production needs."

Those who have been retrained have found automated typesetting easier, both physically and environmentally. The composing room is being rebuilt, the air is cleaner, the lighting is much brighter, and in about three months the entire room will be tiled. Just as at the *New York Post*, some older printers who have chosen not to be retrained have virtually nothing to do. On the whole, however, said one supervisor, "Men can change. They do change. Some kick, but most roll with it."

One further change at the *Daily News* is that the lobster shift has been eliminated. Few were sorry to see it go, although some had become so accustomed to their midnight routine, including driving to work on uncrowded roads and finding a parking space with ease, that they chose the late night shift, from 8:36 p.m. to 3:55 a.m.

The mood at *The New York Times* is the lowest of the three papers. Although it continues to have all three shifts, automation is rapidly shrinking the printers' workload and for many destroying the pleasure and meaning of work. "Boredom is a big problem." "The glamour is gone." "It's miserable" is how some describe *The New York Times* composing room in 1979.

Everyone at *The New York Times* has been retrained, but complete automation has brought with it a number of problems. Only about 15 percent of the type is now set by printers—mostly classifieds and stories written by "stringers," or free-lance writers. Virtually none of the daily paper passes through the composing room:

> At 8:15 in the evening in the old days, you used to see two hundred guys rushing to get out the first edition. No more. One night I looked around the composing room at 8:15 and saw seven guys at terminals—all doing advance work. The edition was in already!

Many of the men literally have no work. How do they spend their time? "Well, there's a patio on the twelfth floor where a guy can sit and get a suntan. We've asked for a TV and Ping-Pong tables." One insightful printer put it this way: "Automation means no work. Management feels it would be unrealistic to try to put [all] these people to work. The men have to realize that this is a problem with no solution. It's a lesson in industrial relations."

Among printers, *The New York Times* is known as a "job shop." By this they mean that the fast pace of the composing room has always been more like a commercial print shop than a daily paper. Ironically, those who chose to work at *The New York Times*, because they liked its work pace combined with its high standards, are being hit hardest by automation. In 1976 there were 700 printers; now there are 500 and although management talks about 225 being about the right number, according to some printers, *The New York Times* will need even fewer. In fact, over and above the usual vacations, about 75 are on productivity leave at all times, an accommodation to there being far too little work to go around.

At least temporarily, the new format of *The New York Times* has saved some work. For decades the daily paper had been in two sections. But during the past three years, it has adopted a four-section format and on a typical day numbers about a hundred pages. *The New York Times* has also added Sunday sections for different suburban communities—Westchester and Connecticut, Long Island, and New Jersey. Circulation is up, and this pleases the men in the composing room, not only because it adds to their sense of job security, but also because *The New York Times* is "their" paper.

The need for fewer printers is only part of the problem at the three papers. Automation is causing friction among those who remain, and its impact on priority is a case in point. In the past, the shift a man worked or his priority was seldom a topic of conversation. A printer could usually change shifts if he wished, and those with low priority could count on higher priority as new men came into the composing room. Now the odds are that a printer will stay on the same shift. Furthermore, with no new men being hired, the one with low priority has last choice on the vacation schedule and in the few remaining ways that priority counts. Also, management now decides on all work assignments, creating more strain. "If you can't cut the mustard, go to the sidelines," one said, growling like a marine sergeant.

The sense of community is further undermined by management's present implementation of the productivity leave. The collective-bargaining agreement stipulated that the leave could be used at management's discretion, but until recently it was generally taken as a six-month lump-sum payment when a printer retired or for "hardship" reasons—a printer's wife was hospitalized, a death in the family. Now, at *The New York Times*, as noted, seventy-five men are on four weeks of paid leave all the time, whether they like it or not. Those with higher priority have the choice summer months, but those with lower priority must take it as their names get posted. The *New York Post* is doing the same, but only in groups of about fifteen, and in the fall of 1979 the *Daily News* began furloughing forty at a time.

The publishers regard this temporary mechanism for reducing the workforce as necessary, but not sufficient, to bring the size of the composing room staff in line with the work to be done. They must respect the 1974 tradeoff of job security for automation, but to speed the printer's exit, they have come up with bigger and bolder incentives. The $2,500 offer in 1974 at the start of the contract is a mere trifle compared to the severance incentives since then. In 1979 the *Daily News* offered $15,000 to be paid either in one lump sum at termination or in sixty monthly installments, and an incentive by the *New York Post* in 1978 was similar. *The New York Times* has made more offers than either the *Daily News* or the *New York Post*, and under the current program of incentives at *The New York Times* a man can leave with as much as $100,000, depending on his age and how much of the productivity leave he has used. All monies paid are subject to the usual withholding for federal, state, and city taxes.

The men refer to the offers by both the *Daily News* and the *New York Post*, and the earlier ones by *The New York Times*, as "found money," but they call the new proposal by *The New York Times* a "buyout." The distinction is linked to age and the eligibility requirements for social security and Medicare. By and large, the preceding offers have attracted only men over sixty years of age, especially those between sixty-two and sixty-five who were planning to retire anyway. The combination of several thousand dollars plus the added protection of medical coverage for a fixed period—for example, the 1979 offer by the *Daily News* stipulated up to three years of medical coverage and the 1978 offer by the *New York Post* offered up to five years—has made it worthwhile for the older man to retire now rather than a few years later. During the last three years, attrition by financial incentives, by death, and by retirement when no dollar incentives were available has reduced the workforce at the *Daily News* and the *New York Post* by some 300 printers and brought it in line, at least for the present, with work assignments at these papers.

The New York Times, however, is not satisfied with the present rate of attrition. It wants many more men to leave, and it is seeking especially to buy out the younger printer. The older one may take it, but for him the dollars are much less. The current incentive is the first to provide step-up dollar amounts for those not yet fifty years old. It offers assistance to help a printer find another job as well as the option to return to *The New York Times* if the new job does not work out. The try-out period may extend as long as one year, and if the pay is less than what a printer is now receiving, *The New York Times* will pay up to one-half his present salary during this period. Management has retained a private consulting group, New Life Personnel Services, to help printers find other jobs. Those who choose self-employment—at a job approved by a New Life Personnel Services counselor—have the same "no-risk try-out," as the offer terms it.

The incentive may be taken in one lump sum or stretched over five or ten years. Health benefits, employer contributions to the pension fund, and life insurance are continued for five years after termination. In addition, any printer who leaves receives the unused portion of his paid productivity leave, which could amount to as much as $12,000.

Further, management wants printers to exit sooner rather than later, and the incentive schedule reflects this desire. For example, a forty-five-year-old who terminates by July 31, 1980, receives $48,000, but if he waits until March 31, 1981, the incentive is reduced to $36,000. And if management decides that a sufficient number have left by July 1980, it reserves the right to cancel the 1981 offer. The announcement ends with this warning: "This is the last offer that we expect to make. I urge you to give it your most serious consideration."

What should the printer do? Take the money and run? Hold on to the security of $450 a week? Should he try to make something of himself elsewhere or sit it out and hope that boredom does not drive him crazy?

From a distance, one might reason that these men have made a good dollar for many years. It is unfortunate that their trade has gone the way of the blacksmith's, but nothing is forever. Policemen and firemen retire after twenty years and go on to something else, why not printers? Besides, many men in similar situations are simply handed a severance envelope, and the boss says, "Good luck." But not printers. Their employer has offered to help them find another job, and to ease the transition, their pay on that job will be supplemented by as much as one-half their pay at *The New York Times* for as long as one year. With this guarantee, a printer could earn more during the try-out period than his regular salary at *The New York Times*. But the issue of a second career for printers is much more complex than a dollar incentive and income guarantees.

Some printers—but only a few—have a sideline that they can convert to a livelihood. Carpentry and landscaping are two examples cited by management. Over the years, a small number have developed outside business interests. For example, one printer owns a junkyard, a second is a wholesaler in stamps, a third sells shirts on the side. But these "moonlighters" are the exception. In recent years, most printers did not work extra hours off the job to meet expenses. Now they will have to decide on a line of work, seek employment possibilities, and possibly even deal with the problem of geographic relocation.

Moreover, it is one thing for a thirty-year-old to start all over again, but quite another for a forty-five-year-old. Very few printers are under thirty-five; most who wrestle with the idea of relocating are well into their forties. They have been in printing their entire working life; they have families to support, children to send to college, a mortgage to pay off; and they worry about inflation. In another job they probably will not have automatic cost-of-living increases, as they do now. These men want to do something with the rest of their lives, but what should they do and where will they find it?

Their concerns and fears are compounded by long-standing ties to their union, the ITU. It has been in their family for a hundred years, and to even think about turning in their union card is painful. "It's like getting a divorce."

A further obstacle besides age and little, if any, experience outside the trade is limited education. Most have no more than a high school diploma, and the printer, for example, who thinks that he might like personnel or industrial relations finds that there is no market for his interest. "I sent out 200 resumes and not one bite" was one man's experience. At this time management will assist a printer who wants to take a course, say, in real estate for certification as a broker, but management is not going to pay for a college education. The emphasis is on trying out another job, not on training for another career or counseling about how to come to terms with it.

For the man over fifty, the problem is more severe. Blue-collar or even white-collar jobs, such as in the growing computer industry, are aimed at younger persons and pay lower wages than printers have been receiving. Selling insurance, real estate, or even men's ties is an option, but success depends on commissions and the potential of the territory to which one is assigned. And these lines of work are not notable for job security, even if they will take the older man. The alternative of remaining in the composing room for printers both under and over fifty is also very difficult—there is boredom, nothing to do, tension among fellow printers, and full realization that their employer no longer needs or wants them on the premises.

At this time neither the printer at the *Daily News* nor the one at the *New York Post* faces the dilemma of the man at *The New York Times*. They will know how he resolves his situation, they will learn from his experience, and at the same time they will hope it does not happen to them.

Virtually all the printers voted for the automation agreement, but few realized how swiftly the changeover from hot metal to cold type would take place and what a surplus of men it would create in so short a time. Management may well have been equally surprised, because according to one informant, "The publishers would not have signed an eleven-year contract, if they had known it [automation] would go so quickly."

Just a few years ago printers thought that the big issue was learning the new processes and helping their employers modernize, so that the papers could be competitive financially and better than ever. They counted on job security for life and voluntary retirement. They gave up their craft, work rules that had been hard-won, and with dignity and determination made the transition to automated work. Their craft, and in particular their job, gave them a sense of property not unlike that which the capitalist values and protects and the professional secures through his educational investment. Fathers and uncles had passed the craft to the men now in the composing room, and although this generation did not expect their children to carry on the tradition, they are jolted by the suddenness of no longer being valued and not even being wanted.

The dollar cost to an employer of excess employees is high, but the human cost to the individual of living with the present tension and uncertain future is also very high. Assisting men with only a high school education as they turn to a second career in their middle years requires preparation, sometimes schooling, counseling for both husbands and wives, career research, and a supportive rather than threatening environment while they make the transition.

Finally, some who plan to remain fear what might happen when the agreement expires. "There will be war in 1984," said one printer. "They [the publishers] will never agree to pay us to sit around until we're ready to retire.'' Others are more confident of the job guarantee, and they dismiss the worriers—''There are always people who paint a black picture.'' Among the more confident are the supervisors whose skills are needed and the machinists who have become technicians.

The union does not play a role in the proposals management presents to the composing room force, and the men do not expect this of their leaders. The hope is that during the next few years their union will merge with one or more other unions for increased strength at the bargaining table, and they count on Bert Powers, their president, to negotiate the next agreement.

Thus, the composing room is in a state of crisis. The conversion to cold type has drastically reduced the workload, and as even more sophisticated equipment is brought into the composing room, the problem of providing work for those who remain will become still more acute. Although the situation is most serious at *The New York Times*, the handwriting is on the wall at both the *Daily News* and the *New York Post*. Management's answer has been to reduce the size of the workforce not by retirement or normal attrition, but by the more expedient method of offering increasingly stronger incentives to push printers out. But the printers are not ready. Most have nowhere to go, and dollar signs confuse the choice. Must technology take so heavy a toll on the U.S. worker?

8 Summary and Conclusions

Technological change at the workplace can be examined from at least four critical perspectives: the capability of the necessary machinery, the readiness of management to make the dollar investment, the impact on a union of the new conditions, or the response of workers themselves. In this book, the workers' experiences and viewpoints about what has been happening to them have been the center of attention.

In our work-oriented society, a man's worth is measured by what he does, and at question is how men—not younger, minimally trained production workers, but older highly skilled members of a prestigious craft—come to terms with gigantic change in the way they earn a day's pay. At what cost and with what concerns do they accept the fact that their competency and their worth as craftsmen no longer have market value?

What these printers accomplished has now been relegated to history. Nobody is going to reward them for having acquired a complex skill the hard way. The fact that they had served long apprenticeships at low wages doing everything, including sweeping floors and emptying cuspidors, in order to have the opportunity to learn a valued craft matters not at all in today's job market. Who would employ a linotyper or a proofreader, however talented, when the work can be done faster and cheaper by the electronic mating of a typewriter and a television screen?

In the excitement of seeing what advanced computer technology can routinely do, it is all too easy to overlook the concerns of men whose skills are no longer in demand. This is why, in fact, printers agreed to become the subject of this book. In their lifetime, the industry they had come to know so well was dramatically different. Recounting their experiences was, for them, one way of gaining perspective on the social change which confronts them.

Not only had these men struggled many years to master a craft, but even after they had gained official entry as journeymen, it was far from easy. There was no promise of promotion or even a guarantee of steady work. A few started their own print shop, but most were glad just to have a job. And by the time that they married and were raising families, the Depression was upon them, and they struggled like so many others to earn a day's pay. Then, too, there was the problem of print shops going out of business or moving to nonunion towns, and after World War II newspapers began merging or disappearing. In one regard, these men were more fortunate than many because they had their union, Big Six, which gave them a sense of solidarity and identity, as well as protection and help in the job market.

Over the years their wages improved, but steady jobs were never plentiful. Then, as early as the 1950s and dramatically in the 1960s, came the threat of the new computer technology. Outside New York, newspapers were automating, photoduplicating processes were being perfected, and printers knew that the days of their craft were numbered. It really did not matter how much they valued their skills, how vast their knowledge, how artistically and creatively they could match and contrast typefaces; machines had been invented which could store all their knowledge and do the work. One might dispute the quality of the product, but machines could get the work out and with a speed that at first was unbelievable. For example, instead of three days to sort the classifieds for the Sunday edition of *The New York Times*, the work could be done in twenty minutes, and with greater accuracy.

There is no need to explain how eager management was to install the computer technology or to state that in the long run automation would mean jobs, because only with automation could the newspapers stay in business. In fact, in 1976 the concern was that the *New York Post* might not catch up with the new technology and, because of this, have to shut down.

The problem was the short run and, more particularly, the men in the composing room. As individuals, they could do little to protect themselves, but with their union as the go-between they had considerable strength. Until recently the union had veto power over the introduction of new equipment. But by the early 1970s when the collective-bargaining agreement at the papers came up for renewal, it was clear that the union had to consent to automation on a grand scale.

The expectation that with automation productivity would be so great allowed for a solution that could be both masterful and democratic. That solution was a guaranteed job for everyone in the three composing rooms from the printer with high priority to the substitute with few rights. The fact that everyone got job security—both the longtime employee and the occasional one—is a tribute to the union's ideology and evidence, of course, that the publishers were determined to begin a crash program of automating the composing room and accordingly were willing to bargain with the union. Although some worries and fears persist—the contract is only paper and any piece of paper can be torn up—many have confidence in their lifetime job security.

Some printers are critical of the contract and take care to keep an eye on management, but as a group, they do not define their employer as an enemy. Quite the contrary. They take pride in the paper they help to produce, and they want to see their employer prosper and their paper purchased by more and more people. In a very real sense, they see it as "their" paper, and they reach for a copy to take home at the end of the shift. Moreover, despite the radical change in equipment and work flow of the composing room, they continue to enjoy a good measure of control over the pace of their work.

They have other advantages as well. Their immediate supervisors are members of the same union, as they have always been, and although some "bark," most treat printers as coworkers, not subordinates. The chapel chairman and officers of the chapel are in the composing room all three shifts, watchful that the job rights of their union brothers are protected. Management officials may enter the composing room—once a strict taboo—but this is not cause for alarm. Overall, printers like the people they work with, judge their work as interesting, and get satisfaction from doing a good job. They are energetic and lively people who care about their union brothers and have a sense of humor about themselves. Smugness or conceit is not acceptable behavior on the floor of the composing room. As one put it, "If a printer gets uppity, we remind him that today's paper is used to wrap tomorrow's fish."

The Automation of the Composing Room

One can do research, write abstract treatises, or engage in polemics about worker alienation, but men who want to maintain their relatively good standard of living and who cannot see themselves finding a comparable job elsewhere can and do accept the reality of social change with dignity and some optimism. The finding which is verified again and again is that these printers have accepted the changeover to automation. Their resourcefulness, their flexibility, and their capability are much in evidence. In no sense can they be characterized as a bitter, alienated workforce or as a group of men resistant to change.

To be sure, some older men wish that automation could have been postponed until after their time, and if given the choice, they will not be retrained. But most printers who have learned the new methods, as well as those yet to do so, are quite willing to change with the times. The majority say the work is physically less demanding and, in fact, describe it as more challenging and interesting than the old way. From the perspective of the response of the men, then, the changeover has been a success. Moreover, age is no barrier to retraining. It is important to underscore this fact, because all too often claims are made that the older worker cannot be retrained and must step aside for the younger man.

Of equal importance, the data call into question the contention that the machinery and the technology of work itself are the major determinants of job satisfaction (or dissatisfaction). Our findings demonstrate that a worker's orientation toward his job can and does mediate between the nature of what he does and how he feels about it. There is the traditionalist, the man who loves printing, who derives satisfaction from doing a job well and who values the camaraderie of the composing room. Then there is the instrumentalist, the man who defines a job as a way to earn a living. Money and security, the extrinsic aspects of work, are what he seeks primarily; the tasks to which he is assigned are secondary. He wants to do a good job, and he likes his fellow workers, but above all he values work for the standard of living it gives him.

Younger men look to the future. That future is automation, and they want to be part of it. The young traditionalist either has found or expects to find intrinsic satisfaction in his new assignments. Electronics is a challenge which he is eager to meet. The young instrumentalist also wants to be part of that future, but to him the future—automation, to be sure—is a secure, well-paying job with good fringe benefits and then a comfortable retirement.

These alternative definitions of the meaning of work hold even for the older man who prefers not to be retrained. The traditionalist says the computer technology destroys the craft of printing; the instrumentalist, that it is not worth it to him or to management to learn the new methods, because he will not be working much longer.

More generally, whatever a man's appraisal of work, the fact that he has a satisfactory explanation is what counts. It gives him a way of coming to terms with his situation, and it contributes to the positive attitude toward automation in the composing room.

But no change as gigantic and total as automating the composing room is without cost. For management it has meant a huge dollar investment and a gamble that the computer technology has the capability and sophistication to process several daily editions of a large, diversified newspaper. Automation has also exacerbated the problems of the union—loss of jobs, an aging and soon to be pensioned membership, and correspondingly the need to organize new shops and, in all likelihood, to merge with one or more unions. And for the printer the changeover has required accommodation, some of it far from easy.

No printer "owns his job," as previously; the old and respected system of job classification is only history, and priority has come to mean less and less. The goal is to train all-purpose compositors who can perform keyboard and nonkeyboard functions. Some of this work is boring, and patterns have already set in. More often than not, the linotyper becomes a clerk-typist, the handman does paste-up, and proofreading, once the most prestigious of jobs, is reduced to scanning computer symbols and doing billing. Ironically, machinists, the low men in the hierarchy of printing, have now been elevated to technicians, and some of the best jobs are theirs. While all the men appreciate the modern, clean working conditions, machinists are most enthusiastic about the transition from hot metal to cold type. Overall, machinists and others trained to become technicians have gained the most from automation, and proofreaders have lost the most.

Further, now that management can—with only minimal restraints—select which men to retrain and has the freedom to assign a man to any job, priority is no longer of much consequence. Priority still governs vacation and slide days, but has little to do with job assignments. And the elimination of the substitute system also reduces the printer's independence, because he no longer can hire someone to take his place for a day or two. Finally, the six-month paid productivity leave built into the contract is generally used at management's rather than the worker's discretion. Barring "hardship" (a family emergency, a personal problem), the printer does not elect when he will take some of or all his leave time. He may apply for it, but management says yes or no, and during the first few years of the contract management has said no except in cases of hardship or when a printer took it as a lump-sum payment on retirement. In 1979, however, the management of all three papers is using the leave time to furlough excess printers. The present pattern is four weeks of paid leave whether a printer wants to take it or not. At *The New York Times* this means seventy-five men away all the time; at the *Daily News*, forty; and at the *New York Post*, about fifteen.

The Option to Work or to Retire

Given the national norm of a fixed age for retirement, the printers employed at the *Daily News, New York Post*, and *The New York Times* enjoy a privilege accorded few workers. Even their union brothers in the commercial branch of the industry (book and job shops) have mandatory retirement—they must retire by sixty-eight. Understandably, the men at the papers highly value their right to decide if and when they will retire.

The inclusion in the 1974 contract of a $2,500 retirement incentive proved to be a poor inducement, if any, to a printer's retiring sooner rather than later. Men who had already made the decision to retire timed the date of their job termination to take advantage of the bonus. Few others gave the idea serious consideration. And the response of the men one year later to a considerably larger dollar incentive to voluntarily terminate their employment had the same effect. Rather than the fifty whom management had hoped would rush to sign up for a second dollar incentive, only nine took it, and all but one were men over sixty.

The reason that the younger printer did not take the incentive monies is largely shaped by the conditions of the job market. The decision would have involved either uprooting his family, if he wanted to use his printing skills, or finding another job with comparable advantages in a different line of work, an unlikely prospect for the middle-aged man with only a high school education, especially in the tight labor market.

When older printers think about voluntarily terminating their employment at a paper, what most mean is that they are going to withdraw from the labor force. Very few expect to do any work for pay, and if they do, it will be part-time, not full-time, work. This means that they are counting on their pension monies to pay their bills and afford them a few luxuries. Accordingly, the eligibility rules for social security and Medicare set the framework for their decision, and their target date is sixty-five, give or take a year or two. Voluntary retirement to them means the luxury of working a short while longer if they wish, and typically this is for a bread-and-butter reason (paying off a college bill or the mortgage on a house).

By and large, therefore, the men who took the bonus were over sixty-five, and their wives were also eligible for social security. Contract negotiations extended over a two-year period, and these men waited to retire in order to capitalize on the expected benefits. The $2,500 bonus, the functional equivalent of a gold watch, was a welcome gift. But for the printer closer to sixty than sixty-five (let alone the one not yet sixty) and for the man who defines work as his life, no bonus of this size could make retirement attractive.

Of special interest when one considers the attraction of work versus retirement is the hypothesis that disaffection or, more strongly, alienation from work is a motivating factor. Again, the account of these printers provides no indication that they are being driven to retirement because they dislike their work, either the new or the old way. And this includes the few men under sixty-five and in good health who took the bonus. These men liked their work as much as those who wanted to work a few more years.

There are, however, some men who simply do not want to retire. This is particularly the case for those over sixty-five, either never married or widowed, who define work as their life and the composing room as their home and community. It gives them a place to go and a chance to be with people they know and like. Without work, their lives have little meaning. Although the work ethic may no longer be a universal credo, these men still live by it.

With the exception of the small proportion of printers who totally reject retirement, it can be said that printers are about equally divided between those who are enthusiastic about the idea and those who approach it with reservation. The latter are men who say that in some ways they like the idea, but then again they are not sure it is for them. That some are hesitant is not surprising. Until recently, the tradition in the industry was to work all one's life, and the men who retired were the first generation of printers to take this step. Ten years ago these men might not have given it a thought. After all, why would a printer with a steady job and in good health quit working? Now, however, they do think seriously about the idea, which suggests that retirement can become an attractive option to men who did not consider it until they were in their later years of employment.

Printers' attitudes toward retirement also suggest that preparing for retirement does not guarantee liking it. Much has been written about the importance of reading about retirement, talking to family and friends, and saving money. Supposedly, if one does these things, his fears and concerns will lessen and he will find the idea attractive. The account of these printers indicates that this generalization is too simple; preretirement acitivities can have either a positive or a negative effect. Men who have doubts about retirement are just as likely as those who are enthusiastic about it to read an article or a book and to talk to family and friends. It is not that the man who is unsure about retirement shuns information, but rather that all men are selective about what they read, whom they talk to, and what they remember. The man who is eager to retire finds favorable articles on the subject; the one who is not yet convinced finds an article or someone who warns him that retirement can be a big mistake. This may seem obvious, but some advocates of preretirement planning have emphasized the importance of thinking and talking about the idea without taking into account how differently people may react when they turn the glossy pages of a retirement pamphlet or talk with a friend.

As to retirement itself, four themes emerge. First, age is no barrier to making it a success. The seventy-five-year-old who decides to retire can be just as happy as the sixty-five-year-old.

Second, there is a close similarity between preretirement and retirement life. Retired printers stay close to what they know and were like in the past. Indeed, what these men were like *before* foretells their adjustment, not whether they move to Florida, buy a trailer or a boat. Differences in social background, interests pursued as workingmen, and orientation toward retirement separate what we have termed the retreatist from the activist retiree. The latter is a man of higher achievement, more integrated into his union, more diversified in his activities, and more socially mobile. He felt pulled to retirement by the opportunities it offered. In contrast, the retreatist, a man of more limited horizons, felt pushed to retire because either he wanted to get away from the routine of work or his wife had been urging him to quit. Now in retirement, he derives less satisfaction from his new role, it takes him longer to adjust, and he experiences a kind of social anomie. He attempts to compensate for these difficulties by reminiscing about his days as a printer. The activist liked being a printer as much as the retreatist, but he concentrates on making the present period of his life meaningful. After two years of retirement, the activist finds it a more rewarding and satisfying experience than the retreatist.

Third, the wives of these men also vary in their appraisal of retirement. Some are happy to center their life on being at home, and others reach beyond the home, deriving satisfaction from going places and doing things. When husbands and wives share the same perspective, adjustment goes more smoothly than when opposites are paired.

Finally, we singled out the involuntary retiree. This is the man who has stopped working because of poor health, either his own or his wife's. His opportunities for pleasure are quite limited, and his need for resources, both money and community services, is considerable.

Aside from this one group of men with special problems, however, how printers and their wives see retirement highlights the strengths and resources which a person can bring to this period of life. Having realized only later in their working lives that retirement was a possibility and being older than most men when they stopped working make their appraisal of retirement especially worthy of attention. Although some have encountered difficulty, their candor in describing their situation will hopefully ease the transition for the next generation of retirees.

The Union and the Contract

The 1974 contract was signed at a time in the union's history when it was (and still is) confronting a number of critical issues brought on by automation and the changing social and economic climate for trade unionism. Although the printer's skills are respected, there is less and less demand for them, and the printer and his union know this well. The consequent problems for the members and leaders of Local 6, the largest of the ITU locals, bring into sharp focus those which its parent union and many other unions are up against.

The ratio of journeymen to pensioners is declining rapidly, and the newspaper contract has formalized this. No new members are to be hired to replace those who terminate, and the power of priority is sharply diminished with the freedom of management to assign workers as it thinks best. Now, the viability of the union depends on conserving its treasury, organizing nonunion workers, and merger. In combination these pressures are enormous, and all require immediate attention. Tensions and problems are not new to trade unionism, but this union faces an especially difficult challenge because of the new technology.

It would be misleading and an oversimplification, however, to say that the printer's union will completely disappear. What is equally clear, though, is that the old, single-craft union of which Local 6 has been an outstanding example is in the process of being transformed.

In brief, the same kind of accommodation the printer is making at the workplace is being made by his union in its internal operations and in its relations to the larger economic and political structure within which it functions. And, just as the printer's orientation toward work, traditional or instrumental, mediates between the new technology of the trade and how he comes to terms with it, the union's ideology influences its approach to meeting the demands for change. The union's ideology is shaped by tradition, by the stance of its leaders, and ultimately by the response of its membership, because every structural change must be submitted to the rank and file in the form of referenda to be accepted or rejected. The process is time-consuming and divisive and strains the membership to strike a balance between self-interest and the interests of one's union brothers. The process is also democratic.

An indication of the union's strength is that on two recent and heated intraunion issues—the fraternal pension and the voting rights of pensioners—a compromise was evolved which met the pressing needs of its two major constituencies. By guaranteeing the most senior members an additional pension which they valued highly, the union reinforced the fraternal bond between journeymen and pensioners; shortly thereafter the pensioners supported a referendum which negated some of their long-standing voting rights.

With the resolution of these two issues, the union must again pull its membership together to address larger problems—signing up new members and, in all likelihood, merging with one or more unions in the printing trade. The odds are against major victories in organizing, and merger requires accepting a different kind of trade unionism—more centralization, more bureaucratization, more reconciliation of diverse interests.

The substance and the rhetoric of fraternalism are changing, and no one denies the impact of the new technology on the union. But the union is a fighter in the face of adversity. Forced by circumstances largely beyond its control, the union is working hard to shape its organization in such a way that it will preserve the core values of trade unionism—worker protection, equality, and dignity.

The Future Unfolds

The eleven-year collective-bargaining agreement was designed to protect both the worker and management during the period of transition to electronic printing. A return in 1979 to the composing rooms of the three papers indicates that printers have fulfilled their part of the agreement. Both younger and older men have proved not only that they are willing to learn the new methods, but also that they have the necessary ability. Moreover, most of those in their middle or late sixties, who said a few years ago that they were going to retire, have carried out their plans.

At the midpoint of the contract, however, management's interests are not well served, and it is responding accordingly. Because the high-speed equipment functions so well, management, especially at *The New York Times*, wants the retrained printer to leave. At this time, the situation is not as severe at the *Daily News* and *New York Post*, but the pattern of not enough work for the composing room force is becoming more and more evident.

As a result, the printer's worth has been priced not unlike merchandise in a store. But the problem cannot be solved by dollars alone. Even though some assistance in relocating is available at *The New York Times*, for example, the task is not easy and printers' concerns and fears go far beyond connecting with another employer. In effect, most are in a double bind. They cannot stay in the composing room because having nothing to do will destroy their sense of self-worth. But deciding what kind of work to do and finding an available job are difficult, if not impossible, for some. A few, of course, will make the transition easily, but many will not. Unlike policemen and firemen who know that they will retire after twenty years of service and begin a second career in their mid-forties, printers had no reason to expect that they should be developing another line of work on their days off. Now they face this situation, and they have only a few short months to work out a solution.

Printers need far more preparation than this to make the transition successfully. In our society we try to give teenagers at least two years of college before asking them to begin a career. It is equally, if not more, important to give mature men access to the range of resources that now exists for assisting career changes. But at present this is not forthcoming for printers.

Concluding Remarks

At various points throughout this book we have considered some of the implications of the findings, but always in the context either of the individual participants or of the two collective participants, union and management. In this regard, the main questions have been: Whose interests are served? In what ways? At what cost? Before concluding, we want to place the findings in the context of broader policy issues which relate to the quality of working life and incentives to retirement, as well as comment on possible directions for future research.

First, our findings support other research which indicates that older workers can be retrained and that the meaning they attach to work can facilitate their accommodation to new conditions at the workplace. Simply stated, alienation from work is not necessarily a concomitant of technological change.

Second, a number of problems emerge when men not only accept a collective-bargaining agreement but actually become partners in carrying it out. The printers voted overwhelmingly in favor of a precedent-setting agreement, and then they facilitated its implementation by undergoing training and by retiring for the most part, once they reached an age when they would be eligible for social security and medical benefits. But the human cost has been high, as it is likely to be in other industries caught by technological change. What has happened, in effect, is that the retrained printers have worked themselves out of jobs.

One should not infer from this statement, however, that workers should never become partners in technological change. Had printers taken this position, in all likelihood one or more papers would have shut down. Rather, assistance to make the transition successful from a human as well as economic standpoint needs far more emphasis. In the case of printers, job security and voluntary retirement were expected to accomplish this purpose, but they have not, given the frightening problem of "nothing to do." Proposals to buy out printers, even at $100,000 per man, are a bargain for management. But the fact of the matter is that it is the printer who pays. Fast dollars may look big, but what is needed is a program to provide realistic options for a second career, skills and knowledge to convert the loss of a job into an opportunity, and respect for the individual pushed aside by advances in technology.

Third, as a group, printers represent that segment of U.S. society that derives immense satisfaction from their social mobility. They have done better than their fathers, their sons have done better than they have, and this means a great deal to them. What they seek is work to do and a wage scale to meet their expenses and give them a few luxuries. Private and public funds invested in making the worker happier might better be directed to implementing alternative ways to develop jobs, especially for the middle-aged person who must find a second career.

In a period of rapid technological change, the question also arises as to how to encourage workers to retire at an earlier age. The lesson to be learned from this group of craftsmen is that no token bonus will tempt them to begin a second career or to retire earlier than they had planned. Most will retire, but they will do so when they can maximize their retirement benefits. The industry that wants to promote early retirement is better advised to improve its pension and health benefits than to develop ad hoc bonus packages or elaborate preretirement programs in the hope that these efforts will induce a worker to retire sooner.

Much of the evidence on retirement in the United States suggests that most workers expect to retire, and in this respect we anticipate that the impact of recent legislation extending the retirement age to seventy will be minimal as long as workers have a benefit structure which permits them to maintain a decent standard of living in retirement and gives them some protection against inflation, including medical expenses. Only a small minority of workers reject the notion of retirement—these men would work just as long as health permits—and it is hoped that their preference can be accommodated in this pluralistic society.

Because the primary focus of this book has been the individual at different points in the work-retirement cycle, we have not directed as much of our attention to the larger institutional interests which play such a dominant role in the decision of who shall work and who shall retire. This is one of the three directions in which we would hope more research will be conducted.

First, there is need for organizational analyses of labor-management relations, most particularly in industries which are being transformed from labor-intensive to capital-intensive. Such studies can provide models for understanding the process by which the competing interests of management and workers are resolved in the context of fewer jobs and management's determination to change the mode of production.

Second, research into the role of trade unionism in a changing industrial setting would broaden our knowledge of the social consequences of technology. In the face of the pressure of such change, what functions do unions serve? In what ways are their functions being redefined and with what impact on economic and political decisions?

Third, the retirement experience of the increasing number of people who occupy this status requires greater attention if we are to know their capabilities, interests, and problems, and how these may change over time. Although it is widely recognized that individuals may spend as much as one-fourth of their lifetimes as retired persons, beyond considerations of money and health we know relatively little about what makes for a successful retirement. We know even less about the special concerns and resources of these people and how to make their lives more meaningful on an individual and a societal level.

Hopefully, information on the experience of work and retirement will be used as a guide for the formulation of public policy to enhance the quality of life of the individual and the wider society.

Appendix A Research Methods and Sampling Procedures

Methods of Data Collection

The study of the effects of the 1974 collective-bargaining agreement between the Publishers' Association of New York City and New York Typographical Union Number 6 employed three distinct but related research methods: informant interviews, participant observation, and a survey.

Informant Interviews

The field work began in the autumn of 1975 with interviews of chapel officers at the three papers and various officials and staff at the headquarters of Local 6. During this period, we also conducted preliminary interviews with the general foreman and shift foremen at each paper. This preparatory work was enormously helpful. Almost an entire year had elapsed between the time we designed the study and when the funds were in hand to begin it. Although from our experience in research this gap in time is not unusual, it does mean that one must catch up with what has been happening in the research setting before formally beginning the work as planned. Visits to the composing room and union headquarters brought us up to date with the issues confronting the men and their union in the changeover to cold type and helped us to learn a new vocabulary so necessary for understanding the automation of printing. Further, our early interviews with union and management officials suggested many items for inclusion in the survey instruments.

Informant interviewing continued throughout the study period, and it was again particularly useful as we began to analyze our survey data. Several times, when a question arose about the meaning of a particular finding, a subsequent interview with a chapel officer or supervisor would yield information that was helpful in interpreting the data.[1] Altogether, counting chapel officers, supervisory personnel, staff members of Local 6, management officials, and personnel in charge of retraining programs, we conducted more than fifty informant interviews.[2] Within a fairly short time we developed sufficient rapport with our informants that it was not unusual to receive a telephone call alerting us to a new development in the automation process or to a referendum or election issue which was on the union docket.

Participant Observation

Participant observation, as Gans (1962) notes, is a generic term for a variety of observational methods. In our case, most of the observational activity was of the type termed *researcher acts as observer* (Gans 1962, p. 338). That is, our primary function was to observe, not to participate. On many occasions between interviews, or on a special trip just for that purpose, we walked around the composing room, spoke with a printer as he set a page or operated a piece of equipment, and watched other printers make up a page, talk about an assignment, or rush to meet the edition deadline. From time to time, we also joined printers for lunch or a coffee break in the cafeteria. In addition, we attended chapel meetings, which are held monthly at each newspaper, meetings of the Big Six Pensioners Club, as well as general membership meetings of Local 6.

The field work proceeded more easily than we had anticipated. We quickly learned that the business of turning out a newspaper has its own predictable rhythm. For example, Thursday and Friday are the busiest times, as the large weekend or Sunday edition is being prepared. As a result, almost no observation or interviewing was done on those days or nights. Edition time varied at each of the papers, and after seeing the excitement and heightened pace of work on several occasions, we made this our time for coffee in the cafeteria. Going to the cafeteria also gave us a chance to fill in our notes and to talk about our hunches and observations.

Sometimes a story would break when we were on the composing room floor, and we, like the printers, were caught up in the speed and urgency of the moment as linotype machines began to clack busily and handmen rushed to revise a front-page headline or story and foremen barked orders. Throughout the field work we were careful to observe and to interview so as not to interfere with the production schedule. We willingly went to the composing room at all hours of the day and night—after midnight on Saturdays, before dawn on Mondays, and times in between. Whenever the next respondent on our list was engaged in a task from which he could not be interrupted, we interviewed another man or waited until someone was available. Never once were we made to feel unwelcome or intrusive in the composing room of any of the papers.

Within a short time, the printers began to take for granted that "the women from Columbia University" might be in the composing room at almost any time of the day or night, stopping to ask a question or two or requesting that they come into an office to be interviewed. Frequently we were met with good-natured gibes or questions such as "Will my picture be in the book?" or "What are you doing out at three o'clock in the morning?"

The mood and pace of the lobster shift are somewhat different from those of the other shifts.[3] The printers on the lobster shift recognize that they "live on the other side of the moon." With no edition deadline to meet, there is more of an air of relaxation, less pressure, no flashing lights to signify the completion of a page for the pressman. Late Saturday nights, especially at one of the papers, were best of all. One of the men brought in cream cheese and hot bagels for everyone.

Finally, the participant observation was facilitated by the wholehearted cooperation of the printers themselves. Most are outgoing and friendly, and they enjoyed talking with us about their many years in the trade. They also were eager to demonstrate the intricacies (or simplicity) of the new equipment they were being trained to operate, proud to show us how the old linotype machines worked, and ready to display their skill at setting up a page. Indeed, the months in the composing room contributed substantially to our understanding of the life and work of the printer. They were also hard work, but very enjoyable.

Survey of Printers, Retirees, and Wives

While the account of the working life of the printers was largely the product of the many hours spent at the papers, the bulk of the data was obtained through the surveys of working printers, retirees, and the wives of both groups. The surveys were conducted during the spring and summer of 1976. A staff of six, together with the authors, interviewed 408 working printers. The interviews were conducted at the workplace (with the consent of the chapel officers and supervisors) and took between thirty and forty minutes each. Retirees were surveyed by mail during this same period, while the 200 wives (150 of working printers and 50 of retirees) were questioned by telephone.

At no time did we pressure a printer for an interview. All but two signed the necessary informed-consent statement prior to the interview.[4] The two who did not were willing to be interviewed but refused "in principle" to initial any sheet of paper. Most printers were not only willing, but impatient for their interview. Altogether, we interviewed one out of every four printers at each of the papers. Some who were not in the sample were clearly disappointed that they had been excluded, and several asked to be interviewed. We agreed, but the data from these interviews are not included in the analysis.

We attribute the almost universal cooperation of the printers (as well as our 94 percent response rate) at least partly to the fact that the interviews took place not in the homes of printers as originally planned, but at the workplace. As long as the supervisor had no objection, they willingly gave the necessary time at work. Further, as a result of the many hours spent "just observing," the printers became accustomed to seeing us in the composing room. This may well have dispelled any lingering reservations some may have had about being interviewed.

The Sample

Working Printers

The first step in preparation for drawing the sample required listing the total number of working printers (the universe) at the three newspapers. This universe consisted of 1,754 printers, and with the cooperation of the union we obtained the following data for each printer:

1. Name[5]
2. Age
3. Date of membership in Local 6
4. Date union card deposited at the newspaper, thus establishing the printer's priority (seniority)
5. Date eligible to retire with full pension benefits

These data for the 1,754 printers in the universe were processed and analyzed. At this point we learned that 211 printers who had been eligible to retire with full benefits during the bonus incentive period had not exercised this option and were still working. We decided to interview two out of three of these older printers to serve as a control group for the retirees, thus enabling an analysis of the factors which motivated one group to retire and a similar one to keep working.

A second comparison which suggested itself from the literature on "problems" of retraining older workers was that between the older and the younger printers. Accordingly, we stratified the universe of printers by age, and after eliminating the category who had been eligible to retire with full benefits (but who had not), we sampled an equal number of men under forty-five and those forty-five or older.[6]

Our aim was to have equivalent numbers of each group to match the number who had retired during the six-month bonus period with full pension benefits. This group consisted of 140 printers. Table A-1 presents the figures on the universe and sample of printers in each of the three categories, together with the sampling ratios used to obtain an approximately equal number of cases in each category.

The sample was drawn separately for each newspaper to ensure that none of the three newspapers would be disproportionately represented. The final sample of printers, by age category and by newspaper, can be seen in table A-2 along with the response rate for each age category.

Table A-1
The Universe and the Sample of Working Printers

	Under 45	*45 and Older*	*Eligible to Retire*	*Total*
Universe	363	1,180	211	1,754
Sample	145	148	141	434
Sampling ratio	.40[a]	.0125[a]	.667	.25

[a]In order to obtain approximately equal numbers of printers in each category, we oversampled those under forty-five years old and undersampled those forty-five and older. This decision was motivated by the desire to have a large enough representation of younger printers to compare their response to automation with that of the older workers.

Retired Printers

According to union records, 273 printers had retired during the bonus period. We chose to study only the 140 who had passed their sixtieth birthday and who were entitled to full benefits from the negotiated pension.[7] Nine of these were women, and we omitted them from the study.[8] By the time the field work started—about one year later—thirteen retirees had died. Of the 118 to whom we mailed questionnaires with accompanying explanatory letters, 87 responded, yielding a response rate of 79 percent.[9]

Wives

At the conclusion of the interview of working printers (and at the end of the questionnaire to retired printers) we asked for the telephone number of those who were married so that we might call their wives to ask a few questions about the woman's perspective on work and retirement. All but a few printers supplied us with this information. We interviewed 200 wives, 50 from each of the four categories: under forty-five, forty-five and older,

Table A-2
Sample of Working Printers

	Under 45	*45 and Older*	*Eligible to Retire*	*Total*
Daily News	76	63	66	205
New York Post	19	23	21	63
The New York Times	50	62	54	166
Total in sample	145	148	141	434
Total interviewed	143	135	130	408
Response rate	99%	91%	92%	94%

eligible to retire, and retirees. The names were randomly selected from a list of all the names provided by the husbands. The interviews were conducted by telephone and took between ten and twenty minutes each. The interview focused on the woman's response to her husband's work, her attitudes toward retirement, and, in the case of wives of retirees, her assessment of her husband's retirement experience.

A Note on the Presentation of the Data

Technically, when the results for the entire sample are presented, the data should be weighted to correct for different sampling proportions. For several reasons, however, we have not done this in the text. First, when we weighted the categories to correct for oversampling the "under forty-five" and undersampling the "forty-five and over" groups, we found little or no difference in the results. Table A-3, for example, presents the distribution of responses, weighted and unweighted, for several of the variables frequently used throughout the analysis.

Second, our primary purpose in the study was not descriptive, but analytical. By weighting our data, we ensure that they are representative of the universe, that is, printers at New York City's three largest daily newspapers. However, we are concerned less with generalizing to the finite universe of printers at these newspapers than with explaining the response of the printers to the changes which are taking place in the composing room. Why have some printers adapted so readily to these changes, while others have not? Why are some opting to sit it out until retirement, while others are eager to learn the new techniques? Why did some take the bonus and others decline it? How does the printer's orientation toward work help us to understand his response to the automation of his craft? To gain such an understanding, we have relied on both a quantitative and a qualitative

Table A-3
Percentage of Unweighted and Weighted Frequencies for Selected Variables

	Unweighted	*Weighted*
Oreintation toward work	(*N*=396)	(*N*=1,631)
Instrumental	38	39
Mixed	30	31
Traditional	32	30
Will retire within contract		
No	53	57
Yes	47	43
Worked with new equipment		
No	39	39
Yes	61	61

approach, drawing sometimes from the printers' own comments and sometimes from their responses to standardized yes-no or multiple-choice items. We have used these two approaches in complementary fashion. For example, once we discovered that the printers' comments about work and automation suggested an instrumental/traditional typology, we proceeded to divide the sample into instrumentalists and traditionalists and then to test

Columbia University in the City of New York | *New York, N.Y. 10025*

BUREAU OF APPLIED SOCIAL RESEARCH — 605 West 115th Street

PRINTERS' STUDY
Spring 1976

As you may know, we're conducting a study of printers at the <u>News</u>, the <u>Post</u>, and the <u>Times</u> for the purpose of writing a book about printers and how they view their work.

The interview we want to conduct with you takes about 20 minutes and then we have a short form for you to fill out. There are no right or wrong answers to <u>any</u> questions in the interview forms. We simply would like to know your opinions and your experiences.

Your name doesn't appear on these forms and it won't be used in any way in reporting the results of this study. Your participation is completely voluntary and this means you need not answer a question if you wish and that you are free to say at any time that you don't wish to be interviewed.

Because Columbia University feels very strongly that anyone who is interviewed should know both the purpose of the interview and that his participation is voluntary, we'd appreciate your initialing this sheet indicating that you've read it.

____
(initials)

Figure A-1. Printers' Study.

the relationship between a printer's orientation toward work and his response to the changes taking place in the composing room.

Finally, we rejected the idea of weighting the data for another reason. As we saw in the course of our two years in the field, the population of printers is rapidly declining. During that time at least 150 retired or died. Within the next few years many more are expected to do the same. There is little point in attempting to generalize our findings to this small and decreasing universe. Rather, our concern has been to capture the dynamics of change at the workplace and of adjustment to retirement of a group of older blue-collar craftsmen. The men in the composing rooms of the three major New York City dailies provided an optimum research site for this purpose.

Notes

1. See Sieber (1973) for a discussion of the interplay between field work and survey analysis.

2. Moreover, we interviewed many of these informants as many as four or five times during the study.

3. The lobster shift runs from approximately midnight to 8:00 a.m. For the derivation of this term see note 29 in chapter 1.

4. For a copy of the informed-consent form which printers were asked to initial, see figure A-1.

5. At this point we discovered that only thirty-eight (2 percent) of the working printers (and nine of the retired) were women. We decided to eliminate them from the universe listing since only about ten—too small a group for analysis—would have been likely to fall into the sample.

6. This age was selected as a cutting point because we wanted to ensure the inclusion of an adequate number of printers who were young enough to preclude the possibility of their retiring before the end of the contract.

7. Most of the remaining 133 were sixty-five or older but had not accumulated the necessary employment credits for full retirement benefits from the negotiated pension.

8. As with the working printers, we felt this number constituted too small a proportion of the retired population for analysis.

9. Eight questionnaires were returned by relatives who explained that a retiree had died or was hospitalized or too ill to respond. Only three retirees returned questionnaires without completing them. We could not locate the remaining twenty. Several questionnaires were returned marked "addressee unknown." Others may never have reached the retiree, and in still other

cases it is probable that the retiree either had died or was too ill to respond with no helpful relative to inform us of this fact. We attempted to locate nonresponding retirees by telephone without success. As we were told by the mailroom clerk in the union benefits office, many retirees travel from one part of the country to another, visiting children or other relatives, and just leave a mailing address to which their monthly benefit checks are sent.

Appendix B Distribution of Responses to Survey Items

The distribution of responses to items asked of working printers, retired printers, and their respective wives is presented in this appendix. The items appear in the approximate order in which they were asked of respondents. Some questions were contingent on a response to a preceding item or were asked only of an appropriate subgroup; in these instances the base upon which the percentages have been calculated is much smaller than the number in the total sample.

For each item, the base upon which the percentages have been calculated appears in parentheses. From this number, we have subtracted those not answering a given item as well as "don't know" responses (unless a "don't know" response was of substantive interest). Rather than imply a precision not warranted by the size of the sample, we have rounded our percentages to the nearest whole number.

The Working Printer	*Percentage of Working Printers*
Years in printing:	
20 or less	24
21-30	33
31-40	16
41 or more	27
	(406)[a]
Proportion of working life in the trade:	
Virtually all	79
About three-fourths	15
One-half or less	6
	(407)
Member of family in printing:	
Yes	44
No	56
	(405)
If 15 and starting over again, would choose:	
Printing	34
Another occupation	66
	(403)
If printing, why?	
Like printing, no reasons specified	44

The Working Printer	*Percentage of Working Printers*
It's creative, interesting, educational	39
Good pay, benefits, working conditions	33
Personal reasons (for example, the only work I know, difficult to enter another line of work)	19
	(137)[b]
If another occupation, why?	
No future in printing	46
Better opportunities in other fields	42
Printing is changing (less interesting or creative)	28
	(271)[b]
Shift:	
Day	32
Night	42
Lobster	26
	(408)
Main things like about the work:	
Interesting	46
Friends, camaraderie	22
Money, benefits, security	22
Just like it, no reasons specified	20
Easy	16
Working conditions, commuting	14
Like the newspaper	9
	(408)[b]
Anything(s) do not like about the work:	
Yes	45
No	55
	(396)
If yes, things do not like:	
Boring, repetitious	35
Supervisors	28
Hours, commuting, neighborhood	13
Not a craft anymore	12
Dangerous, unhealthy, poor equipment	11
Fellow workers	5
	(188)[b]
Any sections in 1974 contract particularly liked:	
Yes	90
No	10
	(390)
If yes, which sections?	
Job security, length of contract	77
Cost-of-living clause	52

The Working Printer	*Percentage of Working Printers*
Productivity leave	38
Management can automate	12
Pension, benefits	11
Substitutes got jobs	7
	(365)[b]
Any sections not pleased with:	
Yes	39
No	61
	(403)
If yes, which sections?	
Hours, vacations, no overtime	34
Management control over automation	27
No new jobs	25
Medical benefits	22
	(161)[b]
Any groups especially benefited from contract:	
Yes	60
No	40
	(385)
If yes, which groups?	
Substitutes (got jobs)	41
Older men (dollar incentive)	41
Younger men (job security)	25
	(249)[b]
Any groups lost out:	
Yes	29
No	71
	(391)
Taken job training course(s) other than at newspapers:	
Yes	16
No	84
	(401)
Worked with the automated equipment:	
Yes	60
No	40
	(406)
A. Printers Who Have Worked with Automation:	
Length of time:	
3 months or less	18
4-6 months	17
7-11 months	8

The Working Printer	*Percentage of Working Printers*
1-2 years	40
More than 2 years	17
	(241)
Found working with automation physically:	
More demanding	18
No difference	24
Less demanding	58
	(245)
Found working with automation intellectually:	
More demanding	53
No difference	21
Less demanding	26
	(246)
Found working with automation:	
More interesting	48
No difference	17
Less interesting	35
	(243)
B. Printers Who Have Not Worked with Automation:	
Would like to:	
Yes	55
No	39
Depends	6
	(153)
Thinks working with automation will be physically:	
More demanding	19
No difference	21
Less demanding	60
	(148)
Thinks working with automation will be intellectually:	
More demanding	51
No difference	21
Less demanding	28
	(148)
Thinks working with automation will be:	
More interesting	51
No difference	13
Less interesting	36
	(148)
C. All Printers:	
Thinks will enjoy working in fully automated composing room:	
More	42

The Working Printer	*Percentage of Working Printers*
Less	23
No difference, depends	27
Can't say, won't be here	8
	(403)
If more, why?	
Will be more interesting, challenging	49
Will be cleaner, safer	36
Will be easier	24
No choice anyway	8
Opportunity to learn something new	8
	(171)[b]
If less, why?	
Will be boring	44
Will not be a craft anymore	26
Like what I'm doing	25
Too old to change	13
	(93)[b]
Rating of ITU (compared with other unions):	
The best	47
Better than most	32
About the same	10
Not as good as most	10
	(400)
Are there things Big Six should be doing that it's not doing?	
Yes	58
No	34
Other	8
	(402)
If yes, what things?	
More pension benefits	41
Organize, merge	37
More training for automation	14
More political action	12
More vacation, 4-day week	10
	(234)[b]
Pensioners should be able to vote on all union matters:	
Yes	44
No	53
Don't know, depends	3
	(405)
How voted on the ITU fraternal pension referendum:	
To continue the fraternal pension	65
To drop it	31

The Working Printer	*Percentage of Working Printers*
Didn't vote	4
	(396)
How voted on the Big Six fraternal pension:	
In favor	57
Against	36
Didn't vote	7
	(405)
Applied for associate membership:	
Yes	43
No	57
	(384)
Party support: International:	
Progressive	53
Independent	9
Neither (votes for man or issue)	38
	(398)
Party support: Local:	
Progressive	57
Independent	7
Neither (votes for man or issue)	36
	(396)
Spend more time off the job with:	
People in other lines of work	83
Mixed	10
Printers	7
	(403)
Time spent with other printers off the job (outside of union meetings):	
Never	40
Less than once a month	26
About once a month	19
More than once a month	14
	(406)
Member of any nonunion club or organization:	
Yes	42
No	58
	(399)
If yes, which organization?	
Fraternal, religious	56
Veterans' group	29
Political, civic	20

The Working Printer	*Percentage of Working Printers*
Sports or hobby group	12
Other	12
	(174)[b]
Attitude toward retirement:	
Looks forward to it	59
Mixed feelings	21
Dislikes the idea	12
Don't know	8
	(407)
If given the choice, would prefer to retire:	
All at once	50
Gradually	43
Don't know	6
	(405)
Attitude toward mandatory retirement:	
Should be a fixed age	32
Should be left to the individual	65
Don't know	3
	(405)
Expects to remain at paper until retirement:	
Definitely	87
Probably	10
No	3
	(403)
Expected age of retirement:	
61 or younger	19
62	25
63-64	7
65	27
66 or older, or never expect to retire	22
	(406)
Expects to retire within the life of the contract:	
Yes	47
No	53
	(367)
If yes: expects to work for pay after retirement:	
Yes	31
No	65
Don't know	4
	(177)
Hobbies or other activities:	
Yes	84

The Working Printer	*Percentage of Working Printers*
No	16
	(177)
Has done any of the following in connection with retirement:	
Started putting aside money	72
Talked with friends or relatives	70
Read anything on retirement	42
Talked with a doctor	20
Attended meetings or classes	4
	(177)[b]
Why decided to continue working although eligible for full pension benefits:	
Like working	39
Waiting for full social security and Medicare benefits	35
Need the money	23
$2500 not enough of an inducement	22
Personal (not ready psychologically, would be bored)	21
	(124)[b]
Rating of pay:	
Excellent	34
Good	57
Fair or poor	10
	(406)
Rating of job security:	
Excellent	53
Good	38
Fair or poor	9
	(406)
Rating of prestige of printing:	
Excellent	23
Good	51
Fair or poor	26
	(406)
Feels job is:	
Really interesting	58
All right	34
Pretty dull	8
	(405)
Work gives a feeling of a job well done:	
Strongly agree	30
Agree	42
Mixed feelings	22
Disagree	4

The Working Printer	*Percentage of Working Printers*
Strongly disagree	2
	(405)
Work is too simple for ability:	
Strongly agree	16
Agree	23
Mixed feelings	26
Disagree	26
Strongly disagree	9
	(403)
Most of the people are nice to work with:	
Strongly agree	35
Agree	50
Mixed feelings	15
Disagree	1
Strongly disagree	—
	(405)
Has been a chapel officer:	
Yes	31
No	69
	(401)
Would like to be a chapel officer (again):	
Yes	13
Don't care	12
No	75
	(359)
Would like to be a foreman:	
Yes	32
Don't care	19
No	49
	(393)
If given a choice, would prefer to be:	
A foreman	34
A chapel officer	11
Either	11
Neither	44
	(397)
Percentage who have done any of the following:	
Attended union social functions	26
	(403)
Participated in union clubs or sports groups	11
	(400)

The Working Printer	*Percentage of Working Printers*
Served on a union committee	6
	(398)
Been an officer or candidate for office in an ITU local	6
	(397)
Perception of the printer's influence on how the local runs things:	
A lot of influence	8
Some influence	48
Not much influence	28
No influence at all	16
	(402)
Asked by other printers for ideas or information on union matters:	
Often	18
Sometimes	49
Not really	33
	(404)
Voted in last ITU election:	
Yes	97
No	1
Don't remember	2
	(403)
Voted in last Local 6 election:	
Yes	98
No	1
Don't remember	1
	(405)
Likely to vote (in coming election):	
A straight ticket	24
A mixed ticket	65
Undecided	11
	(400)
Party membership:	
Progressive	28
Independent	3
Neither	69
	(402)
Attended party meeting:	
Yes	12
No	88
	(403)
Campaigned for party candidate:	
Yes	27

The Working Printer	*Percentage of Working Printers*
No	73
	(402)
Age:	
Under 45	33
45-54	21
55-61	12
62-64	18
65 or older	16
	(408)
Marital status:	
Married	83
Single	9
Divorced, separated	5
Widowed	3
	(408)
Residence:	
Apartment in Big Six Towers	6
Apartment elsewhere	34
Private home	58
Other (hotel, rooming house)	2
	(406)
Rent or own apartment or house:	
Rent	38
Own	62
	(397)
Number in household:	
One	11
Two	30
Three or four	38
Five or more	21
	(401)
Number of children:	
None	4
One	18
Two	43
Three	20
Four or more	15
	(340)
Schooling:	
Some high school, or less	26
High school graduate	48
Some college	22

The Working Printer	*Percentage of Working Printers*
College graduate or more	4
	(393)
Wife's schooling:	
Some high school, or less	19
High school graduate	61
Some college	13
College graduate or more	7
	(344)
Wife's status:	
Working full-time	24
Working part-time	18
Keeping house	50
Retired	4
Other (in school, looking for work)	4
	(335)
Age of wife:	
35 or younger	15
36-40	19
41-50	25
51-60	25
Over 60	16
	(388)
Religious preference:	
Catholic	50
Jewish	30
Protestant	14
Other or none	6
	(399)
Attend religious services:	
More than once a month	36
About once a month	9
Several times a year	28
Once a year or less	15
Never	12
	(393)
Importance of religion:	
Very important	47
Fairly important	37
Not very important	11
Not important at all	5
	(393)
Political party affiliation:	
Democrat	56
Independent	35

The Working Printer	*Percentage of Working Printers*
Republican	9
	(380)
Health:	
Very good	46
Good	42
Fair or poor	12
	(405)
Total annual family income:	
Under $15,000	4
$15,000-$19,999	46
$20,000-$24,999	34
$25,000-$29,999	9
$30,000 or more	7
	(394)
Family background:	
Jewish	27
Italian	20
Irish	16
Northern European	15
American	11
Black or Hispanic	5
Other	5
	(404)
Place of birth:	
United States	99
Outside United States	1
	(352)
Parents' place of birth:	
United States	31
Outside United States	69
	(403)
Job classification:	
Operator	37
Handman	41
Proofreader	9
Machinist	4
Supervisor	9
	(408)
Newspaper:	
Daily News	47
New York Post	15
The New York Times	38
	(408)

The Retired Printer	*Percentage of Retired Printers*
Years in printing:	
Less than 45	21
45-49	23
50-54	36
55 or more	21
	(87)
Member of family in printing:	
Yes	46
No	54
	(87)
Classification:	
Operator	33
Handman	52
Proofreader	11
Machinist	4
	(82)
Shift last worked:	
Day	29
Night	43
Lobster	29
	(84)
Ever served as chapel officer:	
Yes	39
No	61
	(87)
Work gave feeling of a job well done:	
Strongly agree	61
Agree	34
Mixed feelings	1
Disagree	3
Strongly disagree	—
	(87)
Work was too simple for ability:	
Strongly agree	13
Agree	19
Mixed feelings	15
Disagree	30
Strongly disagree	23
	(86)
Most of the people were nice to work with:	
Strongly agree	54
Agree	31
Mixed feelings	15

The Retired Printer	*Percentage of Retired Printers*
Disagree	—
Strongly disagree	—
	(87)
Found the work:	
Really interesting	80
All right	18
Pretty dull	1
	(87)
The work will be more challenging and interesting with automation:	
Strongly agree	23
Agree	15
Mixed feelings	31
Disagree	18
Strongly disagree	12
	(87)
Printing as a craft will be dead with automation:	
Strongly agree	24
Agree	13
Mixed feelings	21
Disagree	23
Strongly disagree	20
	(87)
The work will be cleaner and quieter with automation:	
Strongly agree	38
Agree	46
Mixed feelings	11
Disagree	3
Strongly disagree	3
	(87)
Any high school girl will be able to do the work of a printer with automation:	
Strongly agree	17
Agree	17
Mixed feelings	20
Disagree	24
Strongly disagree	22
	(87)
Attitude toward retirement:	
Looked forward to retiring	56
Had mixed feelings	32
Disliked the idea	11
	(87)
Before retirement, did any of the following:	
Read anything on retirement	53

The Retired Printer	*Percentage of Retired Printers*
Put aside money for retirement	49
Talked to retired people	51
Talked to a doctor	20
	(87)[b]
Plans made for retirement:	
Many	13
Some	24
Few	23
Really no plans at all	40
	(87)
Attitude about mandatory retirement:	
Should be a fixed age	18
Should be left to the individual	82
	(85)
Would have preferred to retire:	
All at once	62
Gradually	38
	(87)
Adjustment to retirement:	
Quite quickly	39
Within a few months	20
About a year	20
Still have not really adjusted	22
	(87)
Importance of the $2500 bonus in decision to retire:	
Very important	42
Somewhat important	37
Of little importance	14
Not important at all	7
	(81)
Percentage for whom each of the following applied in the decision to retire:	
Looked forward to doing things	50
Tired of working	35
Had enough money	38
Wife wanted him to retire	29
Didn't feel well enough to work	28
Children wanted him to retire	18
Wife's health was poor	13
Other (move South, automation)	5
	(84)[b]
Worked at another job since retirement:	
Yes	2

The Retired Printer	*Percentage of Retired Printers*
No	98
	(85)
Talk on phone with family member:	
Rarely or never	13
Every few months	5
Once or twice a month	20
More often than that	62
	(82)
Visit with a family member:	
Rarely or never	14
Every few months	24
Once or twice a month	16
More often than that	46
	(79)
Talk on phone with friends or neighbors:	
Rarely or never	16
Every few months	8
Once or twice a month	22
More often than that	54
	(82)
Visit with friends or neighbors:	
Rarely or never	15
Every few months	9
Once or twice a month	21
More often than that	55
	(78)
Play cards, bingo, chess, etc.:	
Rarely or never	55
Every few months	4
Once or twice a month	14
More often than that	26
	(76)
Go to a club or organization meeting:	
Rarely or never	38
Every few months	24
Once or twice a month	17
More often than that	21
	(78)
Go to a movie, concert, or play:	
Rarely or never	35
Every few months	30
Once or twice a month	16

The Retired Printer	*Percentage of Retired Printers*
More often than that	20
	(81)
Member of Big Six Pensioners Club:	
Yes	55
No	45
	(83)
Member of organization, group, or club:	
Yes	44
No	56
	(86)
If yes, in which ones fairly active:	
Fraternal, religious	46
Veterans'	23
Senior citizens'	18
Political, civic	13
Sports, hobby	10
Social	5
	(39)[b]
Volunteer, charity work:	
Yes	22
No	78
	(86)
Hobbies or activities:	
Yes	77
No	23
	(84)
If yes, what hobbies?	
Gardening, repairing	49
Traveling, visiting	42
Sports	38
Crafts, photography	35
Reading, TV, cards	18
Cultural, educational	12
Clubs, civic, church	2
	(65)[b]
New hobbies since retired:	
Yes	23
No	77
	(83)
In last year, retiree has:	
Attended Big Six pensioners meeting or social event	37
Read Big Six pensioners newspaper	82

The Retired Printer	*Percentage of Retired Printers*
Read *Local 6 Bulletin*	98
Attended general meeting of Local 6	55
	(87)[b]
In better spirits since retirement:	
Strongly agree	24
Agree	29
Mixed feelings	32
Disagree	7
Strongly disagree	7
	(82)
Often wish had not retired:	
Strongly agree	4
Agree	12
Mixed feelings	18
Disagree	26
Strongly disagree	40
	(84)
Often times when can't keep busy:	
Strongly agree	7
Agree	20
Mixed feelings	6
Disagree	23
Strongly disagree	44
	(84)
Wish had retired earlier:	
Strongly agree	18
Agree	9
Mixed feelings	6
Disagree	38
Strongly disagree	29
	(79)
Thinks retirement has been good for him:	
Strongly agree	48
Agree	28
Mixed feelings	14
Disagree	7
Strongly disagree	4
	(84)
Often misses people at the paper:	
Strongly agree	28
Agree	43
Mixed feelings	23
Disagree	5
Strongly disagree	2
	(87)

The Retired Printer	*Percentage of Retired Printers*
Influence of pensioners on Local:	
A lot	16
Some	46
Not much	31
No influence at all	7
	(87)
Is asked for ideas or help on union matter:	
Often	5
Sometimes	26
No, not really	69
	(87)
In last few years before retirement:	
Attended union social functions	45
	(87)
Participated in union-sponsored clubs or sports groups	14
	(85)
Served on union committee or as delegate to convention	10
	(86)
Officer or candidate for office in any ITU local:	
Yes	17
No	83
	(87)
Voted in last ITU election:	
Yes	97
No	1
Don't remember	2
	(87)
Voted in last Local 6 election:	
Yes	97
No	1
Don't remember	2
	(87)
Party support: ITU:	
Progressive	46
Independent	3
Neither (votes for man or issue)	51
	(87)
Party support: Local 6:	
Progressive	53
Independent	2
Neither (votes for man or issue)	45
	(87)

The Retired Printer	*Percentage of Retired Printers*
Rates ITU, compared with other unions:	
The best	41
Better than most	40
About the same	6
Not as good as most	13
	(87)
Thinks there are things Big Six could be doing for pensioners:	
Yes	54
No	46
	(82)
If yes, what things?	
Increase pension	50
Better medical benefits	32
Include pensioners more in union	18
More social or political activities	5
Allow part-time work in printing	2
Other (organize more journeymen, conduct symposium of pensioners and prospective pensioners)	14
	(44)[b]
Spirit of fraternalism is dying in the union:	
Strongly agree	23
Agree	24
Mixed feelings	30
Disagree	19
Strongly disagree	5
	(84)
Present local leadership can keep union strong:	
Strongly agree	47
Agree	37
Mixed feelings	13
Disagree	1
Strongly disagree	2
	(87)
Pensioners should have right to vote on all union matters:	
Strongly agree	62
Agree	16
Mixed feelings	11
Disagree	5
Strongly disagree	6
	(87)
Automation could lead to the death of the union:	
Strongly agree	13
Agree	13

The Retired Printer	*Percentage of Retired Printers*
Mixed feelings	29
Disagree	26
Strongly disagree	20
	(86)
Age:	
Under 65	24
65-69	47
70 or older	29
	(87)
Marital status:	
Married	79
Widowed	9
Divorced or separated	6
Single	6
	(87)
Number of children:	
None	24
One	25
Two	29
Three or more	23
	(80)
Residence:	
Apartment	51
Private house	48
Hotel or rooming house	1
	(83)
Number in household:	
One	14
Two	74
Three or more	12
	(81)
Place of birth:	
United States	85
Outside United States	15
	(80)
Parents' place of birth:	
United States	23
Outside United States	77
	(83)
Religious preference:	
Catholic	44
Jewish	29

The Retired Printer	*Percentage of Retired Printers*
Protestant	23
None	3
	(86)
Attend religious services:	
More than once a month	30
About once a month	6
Several times a year	23
Once a year or less	22
Never	20
	(87)
Importance of religion:	
Very important	41
Fairly important	38
Not very important	14
Not important at all	7
	(87)
Political party affiliation:	
Democrat	58
Independent	27
Republican	15
	(84)
If Independent:	
Identify more with Democratic party	73
Identify more with Republican party	27
	(22)
Ethnicity:	
Irish-American	16
Italian-American	18
Jewish-American	28
American-American	7
German-American	12
Other	18
	(87)
Health when retired:	
Very good	37
Good	35
Fair	16
Poor	12
	(86)
Health now:	
Very good	30
Good	42

The Retired Printer	*Percentage of Retired Printers*
Fair	22
Poor	6
	(86)
Wife's health when retired:	
Very good	32
Good	38
Fair	20
Poor	9
	(65)
Wife's health now:	
Very good	25
Good	45
Fair	24
Poor	6
	(67)
Time sick in bed at home during last 12 months:	
None	73
One week or less	15
Two or three weeks	8
One month or more	4
	(85)
Hospitalized in last 12 months:	
Yes	26
No	74
	(85)
Health problem or disability:	
Yes	48
No	52
	(85)
Schooling:	
Some high school or less	74
High school graduate	17
Some college	8
College graduate or more	1
	(84)
If married, wife's age:	
Under 60	13
60-64	31
65-69	34
70 or older	21
	(67)
Wife's schooling:	
Some high school or less	59

The Retired Printer	*Percentage of Retired Printers*
High school graduate	31
Some college	4
College graduate or more	6
	(68)
Wife's status at time of retirement:	
Keeping house	57
Working	29
Retired	12
Other	1
	(68)
Wife's current status:	
Keeping house	60
Working	15
Retired	25
	(68)
Moved since retired:	
Yes	32
No	68
	(84)
Considering income and expenses, living:	
Better than before retiring	6
About as well	67
Not as well	26
	(83)
Perception of monthly income:	
More than enough to live comfortably	12
Just about enough	71
Not enough to live comfortably	17
	(84)
Source of income (including any from wife):	
Negotiated pension	100
Social security	99
Fraternal pension	93
Interest and dividends	85
Company pension	77
Wages	10
Assistance, gifts	2
	(87)[b]
Total monthly income:	
Under $600	15
$600-$699	21
$700-$799	21

The Retired Printer	*Percentage of Retired Printers*
$800-$899	11
$900-$999	12
$1,000 or more	20
	(81)
Advice to prospective retirees:	
Save money	31
Develop hobbies	24
Keep working	13
Retire as soon as you can	15
Enjoy each day	9
Take it easy—go slow	9
	(87)[b]

Wives of Working and Retired Printers	*Wives of Working Printers*	*Wives of Retired Printers*
Have attended social functions of the union:		
Yes	20	30
No	80	70
	(149)	(50)
Talk on phone or visit with wives of other printers:		
Rarely or never	63	66
Occasionally	23	24
Fairly often	13	10
	(149)	(50)
Anything(s) union should be doing for printers and their families:		
Yes	38	42
No	62	58
	(148)	(50)
If yes, what things?		
Better medical benefits	38	24
Better pension benefits	20	62
Better working conditions	18	—
More social activities	14	24
Better widow benefits	7	10
	(55)[b]	(21)[b]
Think husband finds (found) his work:		
Really interesting	64	76
All right	29	22
Pretty dull	7	2
	(149)	(49)
Attitude about husband's occupation:		
Glad he is in printing	79	[c]

Wives of Working and Retired Printers	*Wives of Working Printers*	*Wives of Retired Printers*
Wish he were in a different occupation	21	
	(143)	
If glad, why?		
He likes printing	42	c
It's a secure job	33	
It's a good field	17	
It's his whole life	10	
	(113)[b]	
If wish a different occupation, why?		
Not a secure job	30	c
Other field more interesting	33	
Other field more money, opportunity	26	
Printing not a normal life	10	
	(30)[b]	
Husband speaks with wife about work:		
Often	58	c
Occasionally	36	
Rarely or never	5	
	(149)	
How wife feels (felt) about husband's retiring:		
Look(ed) forward	59	59
Mixed feelings	23	22
Dislike(d) the idea	10	12
Don't know	8	6
	(149)	(49)
If look(ed) forward, why?		
He can (does) enjoy life	72	14
(Will have) worked long enough	14	17
He has interests to keep him busy	10	—
We can do things together	—	72
	(88)[b]	(29)[b]
If mixed feelings or dislike, why?		
No plans, it will be (is) boring for him	53	47
(Will be) around the house too much	22	24
Difficult time for a man	12	11
Other	12	29
	(49)[b]	(17)[b]
Prefers (preferred) that husband retire:		
All at once	20	65
Gradually	61	29
Don't know	19	6
	(149)	(48)

Wives of Working and Retired Printers	*Wives of Working Printers*	*Wives of Retired Printers*
Attitude about mandatory retirement:		
Should be a fixed age	21	24
Should be left to individual to decide	79	76
	(148)	(48)
Age at which wife would like husband to retire:		
Under 60	16	c
60-64	26	
65-69	31	
70 or older, never	11	
Don't know, up to him	16	
	(145)	
Any plans couple made (is making) for retirement:		
Yes	23	46
No	77	54
	(98)d	(49)
If yes, what plans?		
Buy a home, move	57	56
Save money	26	24
Travel, visit the children or relatives	26	20
	(23)d	(25)
Wife would like husband to work after retirement:		
Yes	56	c
No	40	
Don't know, up to him	4	
	(96)d	
Wife's adjustment to husband's retirement:		
Adjusted very quickly	c	75
Took up to a year		18
Not really adjusted yet		6
		(48)
Problems of adjustment:		
None	c	62
Wife's activities curtailed		21
Adjusting schedule to husband		15
Developing new (joint) activities		11
		(47)
Wife's current status:		
Keeping house	54	46
Working	38	18
Retired	7	36
	(147)	(48)

[a]Numbers in parentheses are those upon which percentages are based.

[b]Percentages may total more than 100 since printers could give more than one answer.

[c]Not asked of this group.

[d]Only asked of wives whose husbands are forty-five or older.

Bibliography

Alt, John. "Beyond Class: The Decline of Industrial Labor and Leisure." *Telos* 27 (Spring 1976):55-80.

Atchley, Robert C. "Retirement and Work Orientation." *Gerontologist* 11, no. 1(1971):29-32.

______ . *The Sociology of Retirement.* New York: Halsted Press, 1976.

Barfield, Richard E., and Morgan, James N. *Early Retirement: The Decision and the Experience and a Second Look.* Ann Arbor: Institute for Social Research, University of Michigan, 1970.

Belbin, E., and Belbin, R.M. "New Careers in Middle Age." In *Middle Age and Aging*, edited by B. Neugarten. Chicago: University of Chicago Press, 1968, pp. 341-346.

Beynon, H., and Blackburn, R.M. *Perceptions of Work: Variations within a Factory.* Cambridge Papers in Sociology, 3. Cambridge: Cambridge University Press, 1972.

Binstock, R.H., and Shanas, E., eds. *Handbook of Aging and the Social Sciences.* New York: Van Nostrand Reinhold Company, 1976.

Blauner, Robert. "Work Satisfaction and Industrial Trends in Modern Society." In *Labor and Trade Unionism*, edited by W. Galenson and S.M. Lipset. New York: John Wiley & Sons, 1960, pp. 339-360.

______ . *Alienation and Freedom: The Factory Worker and His Industry.* Chicago: University of Chicago Press, 1964.

Bond, Kathleen. "Retirement History Study's First Four Years: Work, Health and Living Arrangements." *Social Security Bulletin* 39 (December 1976):3-31.

Bracey, H.E. *In Retirement: Pensioners in Great Britain and the United States.* London: Routledge and Kegan Paul, 1966.

Braverman, Harry. *Labor and Monopoly Capital: The Degradation of Work in the Twentieth Century.* New York: Monthly Review Press, 1974.

Business Week, July 14, 1978, pp. 72-73.

Butler, Robert N. *Why Survive? Being Old in America.* New York: Harper & Row, 1975.

Canon, I.C. "Ideology and Occupational Community: A Study of Compositors." *Sociology* 1, no. 2 (1967):165-185.

Carp, Frances, ed. *Retirement.* New York: Behavioral Publications, Inc., 1972.

Carroll, B. *Job Satisfaction: A Review of the Literature.* Ithaca: New York State School of Industrial and Labor Relations, Key Issue Series, No. 3, 1973.

Cumming, Elaine, and Henry, William E. *Growing Old: The Process of Disengagement.* New York: Basic Books, 1961.

Davis, Harry E. "Pension Provisions Affecting Employment of Older Workers." *Monthly Labor Review* 95 (1973):41-45.

Donahue, Wilma T.; Orbach, Harold L.; and Pollak, Otto. "Retirement: The Emerging Social Pattern." In *Handbook of Social Gerontology: Societal Aspects of Aging*, edited by C. Tibbitts. Chicago: University of Chicago Press, 1960, pp. 330-406.

Executive Council, International Typographical Union. *A Study of the International Typographical Union, 1852-1963*, vol. 1. Colorado: International Typographical Union, 1964.

______. *A Study of the History of the International Typographical Union, 1852-1966*, vol. 2. Colorado: International Typographical Union, 1967.

Faunce, William A. "Automation and the Automobile Worker." *American Sociological Review* 23 (1958):401-407.

Fields, Theron J. "Company-Initiated Early Retirement as a Means of Workforce Control." Abstract. *Industrial Gerontology* 4 (1970):36.

Friedmann, Eugene, and Havighurst, Robert J., eds. *The Meaning of Work and Retirement*. Chicago: University of Chicago Press, 1954.

Friedmann, George. *The Anatomy of Work: Labor, Leisure, and the Implications of Automation*. New York: The Free Press, 1961.

Gans, Herbert J. *The Urban Villagers: Group and Class in the Life of Italian-Americans*. New York: The Free Press, 1962.

Goldthorpe, John H., et al. *The Affluent Worker: Industrial Attitudes and Behavior*. Cambridge Studies in Sociology, 1. Cambridge: Cambridge University Press, 1968.

______. *The Affluent Worker in the Class Structure*. London: Cambridge University Press, 1969, pp. 1-30.

Havighurst, Robert J. "The Nature and Values of Meaningful Free-Time Activity." In *Aging and Leisure*, edited by Robert Kleemeir. New York: Oxford University Press, 1961.

Havighurst, Robert J., et al. *Adjustment to Retirement: A Cross National Study*. Assen, The Netherlands: Koninklijke Van Gorcum & Comp. N.V., 1969.

Hochstim, J.R. "A Critical Comparison of Three Strategies of Collecting Data from Households." *Journal of the American Statistical Association* 62 (1967):976-989.

International Typographical Union. *Typographical Journal* 167 (December 1975), 168 (January 1976), 168 (April 1976).

Jacobson, D. "Willingness to Retire in Relation to Job Strain and Type of Work." *Industrial Gerontology* 13 (1972):65-74.

Jaffe, A.J. "The Retirement Dilemma." *Industrial Gerontology* 14 (1972):1-88.

Kahn, Robert L. "The Meaning of Work: Interpretation and Proposals for Measurement." In *The Human Meaning of Social Change*, edited by A.A. Campbell and P.E. Converse. New York: Basic Books, 1972.

Kelber, Harry, and Schlesinger, Carl. *Union Printers and Controlled Automation*. New York: The Free Press, 1967.

Kerckhoff, Alan C. "Family and Retirement." In *Social Aspects of Aging*, edited by Ida Harper Simpson and John C. McKinney. Durham, N.C.: Duke University Press, 1966, pp. 131-192.

Komarovsky, Mirra. *Blue-Collar Marriage*. New York: Random House, 1967.

Kreps, Juanita M. *Women and the American Economy: A Look to the 1980's*. Englewood Cliffs, N.J.: Prentice-Hall, Inc., 1976.

Lazarsfeld, Paul F. *Qualitative Analysis*. Boston: Allyn and Bacon, 1972.

Lazarsfeld, Paul F.; Berelson, B.; and Gaudet, H. *The People's Choice*, 3d ed. New York: Columbia University Press, 1968.

Lehr, Ursula, and Dreher, G. "Determinants of Attitudes toward Retirement." In *Adjustment to Retirement*, edited by R.J. Havighurst et al. Assen, The Netherlands: Koninklijke Van Gorcum & Comp. N.V., 1969.

Lipset, Seymour M.; Trow, Martin; and Coleman, James S. *Union Democracy*. Glencoe, Ill.: The Free Press, 1956.

Lockwood, D. "Sources of Variation in Working Class Images of Society." *Sociological Review* 14 (1966):249-267.

Maves, Paul B. "Aging, Religion, and the Church." In *Handbook of Social Gerontology*, edited by Clark Tibbitts. Chicago: University of Chicago Press, 1960, pp. 698-749.

Mullan, C., and Gorman, Liam. "Facilitating Adaptation to Change: A Close Study in Retraining Middle-Aged and Older Workers at Aer Lingus." *Industrial Gerontology* 15 (1972):20-39.

Neugarten, Bernice L., and Havighurst, Robert J. "Disengagement Reconsidered in a Crossnational Context." In *Adjustment to Retirement*, edited by R.J. Havighurst et al. Assen, The Netherlands: Koninklijke Van Gorcum & Comp. N.V., 1969.

New York Typographical Union Number 6. *Negotiated Contract*. New York, July 28, 1974.

______ . *Local 6 Bulletin* 76 (June 1974) to 79 (November 1977).

Palmore, Erdman. "Retirement Patterns among Aged Men: Findings of the 1963 Survey of the Aged." *Social Security Bulletin* 27 (August 1964):3-10.

Parnes, Herbert S. et al. *The Pre-Retirement Years: Five Years in the Work Lives of Middle-Aged Men*, vol. 4. Columbus: Center of Human Research, Ohio State University, 1974.

Perlman, Selig, and Taft, Philip. *History of Labor in the United States*. New York: The Macmillan Company, 1935.

Pizzorno, Alessandro. "Individualistic Mobilization in Europe." In *A New Europe*, edited by S.R. Graubard. Boston: Beacon Press, 1967.

Pollman, A. William. "Early Retirement: A Comparison of Poor Health to Other Retirement Factors." *Industrial Gerontology* 26 (1971):41-45.

Rainwater, Lee. "Making the Good Life: Working Class Family and Life Styles." In *A Symposium on Middle America*, edited by Sar A. Levitan. New York: McGraw-Hill, 1971, pp. 204-229.

Reed, Christopher. "How to Make a Magazine." *Harvard Magazine* 79 (1977):64-71.

Richmond, Samuel B. *Statistical Analysis*. New York: Ronald Press, 1964.

Riley, Matilda W., and Foner, Anne. *Aging and Society*, vol. 1. New York: Russell Sage Foundation, 1968.

Roby, Pamela. "The Condition of Women in Blue-Collar, Industrial and Service Jobs." Mimeo, 1976.

Rones, Philip L. "Older Men—The Choice between Work and Retirement." *Monthly Labor Review*, U.S. Department of Labor, November 1978, pp. 3-10.

Rosow, I. "Adjustment of the Normal Aged." In *Processes of Aging*, edited by R.H. Williams, C. Tibbitts, and W. Donahue. New York: Atherton Press, 1963, pp. 189-191.

Rosow, Jerome M., ed. *The Worker and the Job: Coping with Change*. Englewood Cliffs, N.J.: Prentice-Hall, Inc., 1974.

Shanas, Ethel. *Old People in Three Industrial Societies*. New York: Atherton Press, 1968.

______ . "Adjustment to Retirement: Substitution or Accommodation?" In *Retirement*, edited by Frances M. Carp. New York: Behavioral Publications, 1972, pp. 219-244.

Sheldon, Alan; McEwan, Peter J.M.; and Ryser, Carol Pierson. *Retirement: Patterns and Predictions*. Rockville, Md.: National Institute of Mental Health, 1975.

Sheppard, Harold. "Work and Retirement." In *Handbook of Aging and the Social Sciences*, edited by R.H. Binstock and E. Shanas. New York: Van Nostrand Reinhold Company, 1976.

Sheppard, Harold, and Herrick, Neal Q. *Where Have All the Robots Gone? Worker Dissatisfaction in the '70's*. New York: The Free Press, 1972.

Sieber, Sam D. "The Integration of Field Work and Survey Methods." *American Journal of Sociology* 78 (May 1973):1335-1359.

Simpson, Ida Harper, and McKinney, John C., eds. *Social Aspects of Aging*. Durham, N.C.: Duke University Press, 1966.

Slavick, F., and Wolfbein, S.L. "The Evolving Work-Life Pattern." In *Handbook of Social Gerontology*, edited by C. Tibbitts. Chicago: University of Chicago Press, 1960.

Streib, Gordon F., and Schneider, Clement J. *Retirement in American Society*. Ithaca, N.Y.: Cornell University Press, 1972.

Streib, Gordon F., and Thompson, Wayne E. "Personal and Social Adjustment in Retirement." In *New Frontiers of Aging*, edited by Wilma

Donahue and Clark Tibbitts. Ann Arbor: University of Michigan Press, 1957, pp. 180-197.

U.S. Bureau of the Census. *Census of Population: 1950*, vol. 4. Special Reports, p. I, chapter A, *Employment and Personal Characteristics*. Washington: Government Printing Office, 1953, pp. 109-110.

______. *Census of Population: 1970*. Subject Reports, Final Report PC(2)-7A, *Occupational Characteristics*. Washington: Government Printing Office, 1973, pp. 64-66.

Westergaard, J.H. "Sociology: The Myth of Classlessness." In *Ideology in Social Science*, edited by Robin Blackburn. New York: Vintage Books, 1973, pp. 119-164.

Wilensky, Harold L. "Orderly Careers and Social Participation: The Impact of Work History on Social Integration in the Middle Mass." *American Sociological Review* 26 (1961):521-539.

Williams, Richard H.; Tibbitts, Clark; and Donahue, Wilma; eds. *Processes of Aging*, vol. 2. New York: Atherton Press, 1963.

Wiseman, F. "Methodological Bias in Public Opinion Surveys." *Public Opinion Quarterly* 36 (1972):105-108.

Woodward, Joan. *Management and Technology*. London: Department of Scientific and Industrial Research—Problems of Progress in Industry, Her Majesty's Stationery Office, 1958.

Work in America. Report of a Special Task Force to the Secretary of Health, Education, and Welfare. Cambridge, Mass.: The M.I.T. Press, 1973.

Ziesel, Hans. *Say It with Figures*, 5th ed. New York: Harper & Row, 1968.

Index

About the Authors

Theresa F. Rogers received the B.A. from Hood College and the Ph.D. from Columbia University. She is a lecturer in sociology at Barnard College and a senior research associate at the Graduate School of Business, Columbia University. Her research and teaching interests include the city, work, the delivery of health care, and methods of research. She coauthored *Decentralizing City Government* and has written several articles. At present she is conducting a study of the impact of self-help groups on the rehabilitation of cancer patients.

Nathalie S. Friedman is an alumna of Barnard College and received the Ph.D. in sociology from Columbia University. She has been chairwoman of the Sociology Department at Stern College for Women at Yeshiva University and is currently a senior research associate at Columbia's Graduate School of Business and lecturer in sociology at Barnard College. Dr. Friedman has written several articles and coauthored *Decentralizing City Government*. She has conducted studies in education, the city, and work and is currently studying the orientation and adjustment of women to retirement.

